Introduction to Geometrical and Physical Geodesy
FOUNDATIONS OF GEOMATICS

Thomas H. Meyer

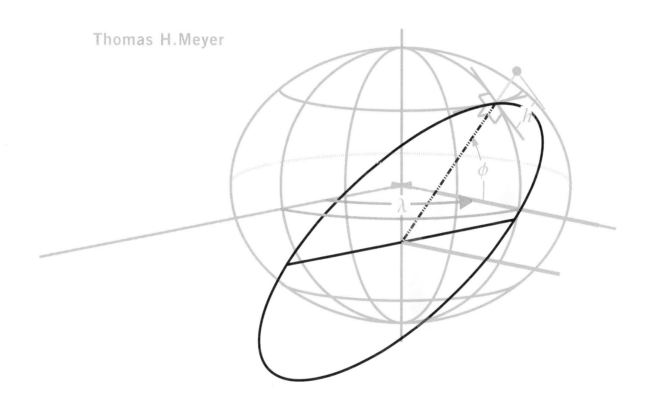

ESRI Press
REDLANDS, CALIFORNIA

ESRI Press, 380 New York Street, Redlands, California 92373-8100
New materials copyright © 2010 ESRI
All rights reserved. First edition 2010
14 13 12 11 10   1 2 3 4 5 6 7 8 9 10

Printed in the United States of America

This work is protected under United States copyright law and the copyright laws of the given countries of origin and applicable international laws, treaties, and/or conventions. No part of this work may be reproduced or transmitted in any form or by any means, electronic or mechanical, including photocopying or recording, or by any information storage or retrieval system, except as expressly permitted in writing by ESRI. All requests should be sent to Attention: Contracts and Legal Services Manager, ESRI, 380 New York Street, Redlands, California 92373-8100, USA.

The information contained in this document is subject to change without notice.

U.S. Government Restricted/Limited Rights: Any software, documentation, and/or data delivered hereunder is subject to the terms of the License Agreement. In no event shall the U.S. Government acquire greater than restricted/limited rights. At a minimum, use, duplication, or disclosure by the U.S. Government is subject to restrictions as set forth in FAR §52.227-14 Alternates I, II, and III (JUN 1987); FAR §52.227-19 (JUN 1987) and/or FAR §12.211/12.212 (Commercial Technical Data/Computer Software); and DFARS §252.227-7015 (NOV 1995) (Technical Data) and/or DFARS §227.7202 (Computer Software), as applicable. Contractor/Manufacturer is ESRI, 380 New York Street, Redlands, California 92373-8100, USA.

ESRI, the ESRI Press logo, @esri.com, and www.esri.com are trademarks, registered trademarks, or service marks of ESRI in the United States, the European Community, or certain other jurisdictions. Other companies and products mentioned herein are trademarks or registered trademarks of their respective trademark owners.

Ask for ESRI Press titles at your local bookstore or order by calling 800-447-9778, or shop online at www.esri.com/esripress. Outside the United States, contact your local ESRI distributor or shop online at www.eurospanbookstore.com/ESRI.

ESRI Press titles are distributed to the trade by the following:

In North America:
Ingram Publisher Services
Toll-free telephone: 800-648-3104
Toll-free fax: 800-838-1149
E-mail: customerservice@ingrampublisherservices.com

In the United Kingdom, Europe, Middle East and Africa, Asia, and Australia:
Eurospan Group
3 Henrietta Street
London WC2E 8LU
United Kingdom
Telephone: 44(0) 1767 604972
Fax: 44(0) 1767 601640
E-mail: eurospan@turpin-distribution.com

Some of the material in chapters 4–6 and 8 first appeared in Meyer 2002.

Chapters 9–13 are based on materials from Meyer et al. 2005a, 2005b, 2006a, and 2006b.

Contents

Preface ix

I Basic Concepts and Tools 1

1 Introduction to Geodesy 3
1.1 Everyday Geodesy 4
1.2 Why Learn about Geodesy? 6

2 Units 9
2.1 Angular Units 10
2.2 Linear Units 14
2.3 Area 15
2.4 Conversions 16
2.5 *Your Turn* 18

3 Reductions and Computations for Plane Surveying 19
3.1 Absolute versus Relative Positions 19
3.2 Linear Distances 21
3.3 Plane Angles 22
3.4 Mathematical Tools 24
3.5 The Inverse Problem in the Plane 28
3.6 Reductions for Plane Surveying 34
3.7 The Direct Problem in the Plane 40
3.8 Sampling Transects 45
3.9 *Your Turn* 47

II Geometrical Geodesy 53

4 Geographic Coordinates and Reference Ellipsoids 55
4.1 The Need for Geodetic Surveying 55
4.2 Ellipses and Ellipsoids of Revolution 57
4.3 Latitude and Longitude 58
4.4 Reference Ellipsoids 61
4.5 Types of Latitude 71

    4.6 Your Turn 72

5 Geodetic Coordinate Systems   73
    5.1 Earth-Centered, Earth-Fixed Geocentric Cartesian (XYZ) 73
    5.2 Geodetic Longitude and Latitude, and Ellipsoid Height (LBH) 76
    5.3 Local Horizontal Coordinate Systems: East-North-Up (ENU) 81
    5.4 Reference Frames and Geodetic Datums 85
    5.5 Frame Transformation Formulæ 89
    5.6 Avoiding Pitfalls 98
    5.7 Your Turn 100

6 Distances   103
    6.1 Types of Distances 105
    6.2 Distance Reductions 118
    6.3 Your Turn 128

7 Angles and Point Positioning   129
    7.1 North and South 129
    7.2 Spherical Trigonometry 137
    7.3 Positioning on a Sphere 139
    7.4 Grid Angles 141
    7.5 Your Turn 143

8 Map Projections   145
    8.1 Developable Surfaces 145
    8.2 Map Projection Types 152
    8.3 Projection Parameters 156
    8.4 Grid Coordinates 160
    8.5 Map Projection Systems 164
    8.6 Your Turn 168

III Physical Geodesy   169

9 Gravity, Geopotential, and the Geoid   171
    9.1 Mean Sea Level 171
    9.2 Physics 174
    9.3 The Geoid 181
    9.4 Geoid Heights 185
    9.5 Geopotential Numbers 188
    9.6 Your Turn 189

10 Height Systems   191
    10.1 Types of Heights 191
    10.2 Height Systems 201
    10.3 U.S. National Vertical Datums 203
    10.4 Your Turn 207

11 Tides — 211
11.1 Tidal Gravitational Attraction and Potential — 213
11.2 Ocean Tides and Body Tides — 218
11.3 *Your Turn* — 218

IV Appendixes — 221

A Vector Algebra and Linear Algebra — 223
A.1 Definitions and Arithmetic Operations — 223
A.2 Vectors and Matrices — 225
A.3 Applications — 227

B Spherical Trigonometric Identities — 229

Bibliography — 231

Index — 243

Preface

I was first exposed to geodesy while studying geomatics at Texas A&M University. I was taking a course in geographic information systems (GIS) and, during an exercise, came upon a dialog box with a field whose value indicated the datum to which my data referred. I had no idea what this meant, so I asked another student. I was told, "Who knows? Just accept the defaults." Unsatisfied, I began to search for the answer to what was seemingly a very simple question. This book is a result of my journey along the fascinating and wonderful trails that question propelled me.

I now teach geomatics at the graduate and undergraduate levels, and I find my question as pertinent today as it was when I asked it as a student – probably more so. The field of geomatics has blossomed into a multibillion-dollar-per-year industry being supported by university programs around the world. All geomatics practitioners need to know some geodesy in order to understand their spatial data and to use them properly. In searching for a text with which to teach my own classes, I felt that the other geodesy books were either at too high a level or too low. My goal for this book is to introduce the concepts of geometrical and physical geodesy at a scope and at a level that geomaticians are likely to encounter in their practice.

My students study natural resource management in the Department of Natural Resources and the Environment at the University of Connecticut. Natural resource managers need to know how to use a GIS, remote sensing tools, and land surveying tools, so I wrote this book with that audience in mind. As a consequence, this book can be useful to a broad spectrum of readers: from those who are specializing in one field of geomatics to those, like my students, who need to know something about them all. Geodesy sits squarely at the intersection of GIS, remote sensing, and land surveying, so it was appropriate to create a broad-based book.

My students receive training on how to solve environmental problems, and some of the problem-solving methods have an engineering flavor. Therefore, this book contains many useful formulæ. In general, geodesy is a mathematically oriented field. Geometry and physics are the foundations of geodesy, and a working, practical knowledge of the subject necessarily requires understanding a fair number of mathematical formulæ. Most of the topics depend only on algebra and trigonometry, but the explanations of gravity and of map projections rely on a little calculus. I have tried to make the mathematics as accessible as I can by providing discussions illustrated by solved examples for almost all of the formulæ in the book. Each chapter includes a *Your Turn* section at the end, offering the reader an opportunity to exercise their newly acquired skills and knowledge. A good working knowledge of algebra and trigonometry will be needed to solve the problems.

This book's treatment of both geometrical and physical geodesy is somewhat unusual; most geodesy books focus strongly on one topic or the other. However, heights play a central role in many modern mapping projects, and geomaticians need an understanding of them as much as they need an understanding of map projection coordinates. The geometric geodesy portion is heavily laced with mathematical formulæ, but these formulæ are essentially algebraic in nature and, therefore, not unduly demanding. However, gravity is the subject of physical geodesy, and the mathematics of gravity begin with differential equations and go deeper from there. In keeping with the level of mathematical accessibility intended for this book, the *Your Turn* problems in the physical geodesy chapters are more conceptual than computational.

The book is divided into three major parts: (I) Basic Concepts and Tools, (II) Geometrical Geodesy, and (III) Physical Geodesy. The introductory discussion of why studying geodesy is relevant for geomaticians in general is followed by a chapter on units of measure, which are numerous, important, and surprisingly subtle. Chapter 3 introduces the concept of reductions. Distances, angles, and other quantities of interest can have various forms. Reductions algebraically transform these quantities among their forms. This chapter also introduces the important direct and inverse problems, which form the basis of positioning. Chapter 4 presents geographic coordinates, which are the coordinates of choice in geodesy, and reference ellipsoids, which are the mathematical surfaces upon which geodetic computations are defined. Chapter 5 answers my question about datums and also presents the major coordinate systems of geometrical geodesy, which, in my view, are the core of geometrical geodesy. Chapter 6 presents many of the various types of distances that arise in various coordinate systems, and chapter 7 explains their various types of angles. Chapter 8 introduces map projections and the grid coordinate systems that are built using them. Chapter 9 is the first physical geodesy chapter, giving an introduction to gravitation, gravity, geopotential, and the geoid. Chapter 10 introduces how height systems are defined and how they are organized into vertical datums. Chapter 11 gives an introduction to how tides affect heights.

An explanation of geometry-based concepts usually benefits from illustrations – much more so when the subject is three dimensional. I have included many figures in this book, which hopefully will be both engaging and illuminating.

The creation of a work such as this is seldom the product of only one person. I am indebted to many people, including dozens of students who used these materials as course notes, and several reviewers who took the time and effort to help me clarify, organize, and present this material more effectively. I specifically want to thank my wife, Ann, for her love and support and her excellent organizational skills.

Tom Meyer
May 2010

Part I

Basic Concepts and Tools

Chapter 1

Introduction to Geodesy

According to the Geodetic Glossary (NGS 2009), **geodesy** can be defined as "The science concerned with determining the size and shape of the Earth" or "The science that locates positions on the Earth and determines the Earth's gravity field." It might be hard to see anything too interesting or exciting in these definitions. After all, generally speaking, the Earth is round, a fact known at least since the time of the ancient Greeks. The earliest known definitive work on this was done by Aristarchus of Samos (ca. 310-230 B.C.E.), an astronomer who worked out a geometric method to determine ratios of the distance between the Sun and the Earth with the distance between the Moon and the Earth, a method that required knowing that all three bodies were round (Maor 1998, p. 63). But what is not obvious from the definition of geodesy is that much modern technology is possible only due to geodesy. For example, without modern geodesy, navigation might still be done with ancient methods (e.g., don't lose sight of the shore) and high-accuracy maps of moderate or larger regions could not exist. Perhaps surprisingly, geodesy plays a central role in a wide variety of cutting-edge sciences, such as geophysics, astronomy, and climatology. The Earth's warming due to climate change is expected to produce rising sea levels in many places, and sea level rise cannot be observed and monitored without geodesy. The observable change in sea level is due to a combination of a change in the ocean's mean surface level and any subsidence of the shore. Geodesy is necessary to tease these two apart. Global warming is also generally reducing the size of ice sheets and glaciers as well as dramatically increasing how quickly they move. Space-age positioning methods make it possible to determine not only the location of an ice sheet, but also its velocity, shape, and size. Geodesy is also a key component of geophysics by making it possible to observe tectonic plate motion, for example.

Geodesy is usually subdivided into geometrical geodesy, physical geodesy, and satellite geodesy, although additional subdivisions are recognized as well. **Geometrical geodesy** is concerned with describing locations in terms of geometry. Consequently, coordinate systems are one of the primary products of geometrical geodesy. **Physical geodesy** is concerned with determining the Earth's gravity field, which is necessary for establishing heights. **Satellite geodesy** is concerned with using orbiting satellites to obtain data for geodetic purposes.

To understand the role of geodesy, it might be useful to look at how maps are often drawn. A **map** is an abstract, scaled, two-dimensional image of some region typically on the Earth's surface, showing the features of interest in their correct relative locations, sizes, and orientations. According to this definition, a globe is not a map because it is not two-dimensional. Neither are directions to the neighborhood gas station drawn on a napkin for a friend. Although entirely adequate for its purpose, such a rendering is a sketch, not a map, because, while schematically correct, it is not an accurate rendition of the features. Maps are drawn essentially the same way whether they are drawn on paper or on a computer screen. The process begins with someone determining positions for all the features of interest. A **position** is (usually) a pair or triplet of numbers, called **coordinates**. A **location** is a place in the real world, whose spatial placement is described by a position. A **coordinate system** specifies how coordinates are assigned to locations.

In order to draw, say, a building on a map, we need to determine the positions of its defining points, like its corners. Linear coordinates, such as the familiar x, y, z, are descriptions of offsets from some point of reference, called an **origin**, each in the direction of its axis. Suppose one corner of the building was found to be 150.25 meters (m) east, 95.51 m north, and 10.59 m above the origin. The coordinates are $x = 150.25$, $y = 95.51$ and $z = 10.59$, and the position is written $(150.25, 95.51, 10.59)$. After the positions have been determined, appropriately scaled symbols of the locations can be drawn on the paper to produce a map (see section 3.5.1 on page 28). The building's edges can then be drawn by "connecting the dots." This example used a **Cartesian coordinate system**. Cartesian coordinate systems have an origin at the intersection of several straight lines, the axes, that are mutually perpendicular.

Linear coordinates are distances as opposed to locations on a number line. Some might ask, "Don't numbers on a number line always represent distances?" The answer is "not necessarily." In a Cartesian coordinate system they represent distances, but in other types of coordinate systems they do not. To illustrate this point, think about street addresses. Street addresses are really just symbols, same as numbers on a number line. They don't *necessarily* mean anything spatial. In Tokyo, street addresses were assigned sequentially in time, not in space, so an address in Tokyo indicates *when* the building was built but not *where* it is located! In contrast, a Cartesian coordinate represents a linear distance from an origin in a specified direction. Geodesy uses many types of coordinate systems, including ones whose coordinates are *angles*, such as latitude and longitude, rather than linear distances. However, regardless of whether a coordinate is an angle or a linear distance, it is always a separation from an origin in a particular direction.

But where do coordinates come from? What coordinate system are they in? Is there more than one coordinate system of a particular type? In the above example with the building, we assumed that it was possible to determine in what direction the building's corner lay from the origin. But how is that possible? Who decides in what direction is north and, once that has been decided, how do we determine other directions for ourselves? By now you may have guessed that geodesy provides the answers to these questions. Some of the primary products of geometrical geodesy are coordinate systems within which realistic geospatial coordinates can be derived.

1.1 Everyday Geodesy

Until recently, most people had not even heard of geodesy and probably even fewer cared. Up to the end of the 20th century, geodesists worked more-or-less quietly behind the scenes, producing the spatial frameworks used by surveyors and cartographers, enabling them to do things like delineate property, design roads, and produce road maps. Things went along as they always had until two technologies changed geodesy's background role: geographic information systems (GIS) and global navigation satellite systems (GNSS). The U.S. NAVSTAR Global Positioning System (GPS) is the latest U.S. GNSS.

Before GNSS, high-accuracy determinations of latitude and longitude were very hard to come by;

1.1. EVERYDAY GEODESY

Figure 1.1: A total station is a theodolite and an electromagnetic distance measurement (EDM) instrument packaged together.

only trained mapping professionals and geodesists knew how to do it, and it was a time-consuming process. GNSS removed this obstacle. GNSS has made it possible for anyone to determine the latitude and longitude of (almost) anywhere on Earth, typically in a matter of seconds, simply by turning on an inexpensive instrument the size of a pocket calculator. This single fact constituted a cartographic revolution and was utterly unimaginable before the advent of the space age, radio communications, the high-speed digital computer, and the atomic clock.

Although technologically marvelous, the ability to rapidly determine latitude and longitude coordinates, in itself, is probably not much more than a novelty to most people. After all, few people need or want to know the latitude and longitude of some place they are trying to navigate to; latitude and longitude are usually not helpful to most people trying to get somewhere. Car navigation usually depends on road maps or written descriptions; knowing the coordinates of the destination is practically useless to a typical driver. However, GIS changed this, too. If a car is equipped with a GNSS and that GNSS communicates with a computer with access to a database of digital maps all compiled in latitude and longitude, then search algorithms can instantly solve routing problems and tell the driver the best way to get to a destination.

Before GIS, if a map of a private property parcel (**plat**) was compiled in any coordinate system at all, it was usually compiled in a **local coordinate system** (see chapter 3). The coordinates of the origin and the orientation of local maps are arbitrary, so local coordinate system maps convey only relative information of their features, such as their separation and size, but not their location in a global sense. Such maps are readily compiled from measurements made using instruments such as total stations (see Fig. 1.1 on the previous page). A total station is a combination of two instruments, one that measures horizontal and vertical angles, called a **theodolite**, and an electronic distance measuring (EDM) instrument. Total stations cannot determine their own global location or orientation. Plats in local coordinate systems are entirely acceptable to landowners and tax assessors who use these maps individually. However, to see how, say, a city's parcels fit together, then local coordinate system maps must be compiled at the same scale, which is not always the case, and the plats would have to be manually fit together like a jigsaw puzzle. The coordinates of such maps are not useful in figuring out which parcels abut one another because it would be an extraordinary coincidence if two side-by-side local coordinate system maps happened to be in commensurate coordinate systems. GIS solved these problems. A GIS can display a map at almost any scale, rescaling as desired. A GIS will also automatically display independently compiled maps in their correct locations (up to the accuracy of the data) if they were compiled in the same coordinate system. The desirability of having all a city's parcels being available for display and analysis as a whole provided the economic impetus to create maps compiled in global coordinate systems, which are a product of geodesy.

Nowadays, whether someone is aware of it or not, geodesy is having an impact on our daily lives. Emergency responders are beginning to navigate using GNSS. Travelers renting cars in unfamiliar cities can navigate like natives thanks to the GNSS navigation system in rental cars. Many people use Web-based mapping services to produce maps with step-by-step directions from their door to their destination, on demand. Ships can sail faster in inclement weather and ride lower in the water without running aground thanks to GNSS and high-accuracy navigation channel surveys. Aircraft can operate in all visibility conditions, and pilots can land with confidence that they will touch down on the runway and not hit any obstacles during their descent. Cell phones can broadcast coordinates in an emergency. Geophysicists are determining the inner structure of the Earth using GNSS both to measure the motion of the tectonic plates and to, essentially, perform computed axial tomography (CAT) scans of its interior. GIS and GNSS are central to a broad range of technological applications, both mundane and exotic, and none of these applications would be possible without geodesy.

1.2 Why Learn about Geodesy?

The GIS revolution created a new generation of cartographers who often are expert computer users but not necessarily trained in the concepts underlying spatial data or the processes by which spatial data are created. This can lead to confusion and errors. Furthermore, GIS analyses always depend upon map projections of the latitude and longitude data from which they were compiled. All map projections distort one or more of length, area, shape and direction, so any analysis that ignores these distortions is less accurate than it could be because, with the knowledge of geodesy, these distortions can be computed and then either eliminated or reduced to negligible levels. Geomaticists need to understand that all geospatial data contain error, either from their measurement or from distortions introduced by geospatial manipulations.

There are many kinds of geospatial coordinate systems and many kinds of map projections. Geospatial data that refer to different coordinate systems cannot be mixed together in a meaningful

1.2. WHY LEARN ABOUT GEODESY?

way any more than temperatures in degrees Fahrenheit can be used with temperatures in degrees Celsius. Spatial data are routinely collected in many different coordinate systems, and using them together in a naïve way results in unrealistic maps. Knowledge of geodesy and cartography is necessary to solve this problem correctly. There are tools to transform the data so that they are all in the same coordinate system. With these tools and the proper knowledge, the geomaticist is free to mix and match data with an understanding of the compromises and trade-offs inherent in each decision made along the way. In extreme cases, ignorance of these distortions and other geodetic topics could lead to maps that could endanger public safety. It's entirely possible, for example, to depict a fire hydrant in the wrong place on a map by not using the proper units for its position. Such a mistake might hinder firefighters in locating the hydrant in an emergency. Gunther Greulich, past president of the American Congress on Surveying and Mapping, stated somewhat facetiously that, "Geography without geodesy is a felony," which underlines the importance of geodesy for all practitioners of geomatics.

1.2.1 The need for flatness

Cartographic products are flat, whether they are paper maps or images on a computer monitor. Maps are flat for many reasons, some rooted in convenience and others stemming from psychology. Humans tend to think of distances as being straight-line distances. We are comfortable with the idea of a horizontal distance and a vertical distance, and we know that they are measured with, say, a measuring tape pulled taught in a straight line. We tend to think of our environment as being flat, and we are very comfortable with (and in fact we prefer), depictions of space that conform to that perception.

The convenience of a flat model of space is clear: globes don't fit in the glove box of a car, and they would have to be very big to show important details of small regions. We usually want to determine distances by laying a ruler on the map and multiplying the measured distance by a (constant) map scale factor. In fact, for distances measured with a ruler on a map to be valid in all circumstances *requires* that the Earth actually be flat.

1.2.2 Round realities

Regardless of our perceptions, the Earth is round. Perhaps the most apparent contradiction of the flat-Earth hypothesis comes from watching ships as they approach land. If the Earth were flat, then ships would suddenly pop into view and be visible from their waterline to the top of their masts all at once. This does not happen. The tops of the masts come into view first with the hulls appearing last. Sailors describe ships as being "hull down" if their masts are visible but the hull is below the horizon. Of course, images of the Earth seen from space are pretty compelling, too.

For the purposes of making maps can't we just pretend that the Earth is flat? Could we, for example, just treat latitude and longitude coordinates, which are angles, as if they were Cartesian coordinates, which are linear distances? For that to work, geographic coordinates would need to have the same properties as Cartesian coordinates, but they do not. For example, every point in a plane has a unique pair of Cartesian coordinates (x, y). But what are the coordinates of the North Pole? Its latitude is 90°N, but what is its longitude? All meridians converge at the poles (Fig. 1.2), so the North Pole apparently is on *all* of the meridians at the same time, meaning any longitude from 0° to just less than 360° seems equally correct. Unlike Cartesian (x, y) coordinates on a plane, longitude and latitude for points on a sphere are not unique in some cases.

Figure 1.2: The orthographic map projection portrays the Earth as it would be seen from space. Meridians of longitude are shown converging towards the North Pole with parallels of latitude running perpendicular to the meridians (data courtesy of NGA).

A map is a planar depiction of a round reality and it's a mathematical fact that such a depiction cannot be free of distortion. Planes are fundamentally different than spheres. It is impossible to concoct some magical mathematical formula to convert one to the other without distorting their geometrical relationships in a significant way. It might be said that the role of geometric geodesy is to bring the world of the round and the world of the flat together.

Chapter 2

Units

Measurement is not possible without units. One might argue that two objects of different lengths could be measured simply by the ratio of their lengths so no units are needed. However, upon choosing either of the objects to be the standard against which the other is compared, the standard has become a de facto unit of length and, thus, units come into being.

People have searched for "good" units throughout history. A unit should be unambiguous and stable so that its realization does not depend on time or place and be universally accessible so that anyone, anywhere, anytime can create an exemplar for that unit. In fact, this notion of universally accessible units was considered a matter of interracial and international equality by some of the leaders of the French revolution (Alder 2002). Units typically have been defined in terms of things that are very consistent or, at least, perceived to not change. For example, one definition of an inch was the width of an adult thumb at the base of the nail (Whitelaw 2007). In some sense, this is a reasonably good definition. The width of a thumb is almost universally accessible because virtually everyone has a thumb, and adult thumb widths are, in fact, fairly consistent at a given time and place. However, three problems arise to cloud the issue. First, there is more than one definition of the inch, including an inch is 1/12 of a foot and that an inch is the length of three barleycorns (ibid.). This multiplicity of definitions is very common. In fact, it seems that almost every unit has had more than one definition and this leads to measurements with very similar unit names, or even identical names, that mean different things. Second, under closer scrutiny, it is clear that these definitions of the inch are actually imbued with a fair amount of intrinsic variability. It is almost inevitable that different exemplars of a unit defined by, for example, the size of a body part, will differ because no two people are shaped exactly the same. Third, it is not obvious how to define many units. Units of length have been based on vague notions such as the distance a man could throw objects, including stones and axes. The distance the king could throw an axe was a legal unit of length in medieval Europe and was still in use in colonial America (Moffitt and Bossler 1998, p. 670). These complications have given rise to a plethora of units for lengths and angles, which complicates the practice of geomatics.

This chapter introduces many of the units commonly used with geospatial data that are widely available from governmental sources and from large geospatial data clearinghouses. There are too many local and arcane units to account for all of them. For example, the Harvard Bridge across the Charles River in Boston has its own unit of measurement, the Smoot, which is based on the height of an MIT student, Oliver R. Smoot. Smoot's body supposedly was flipped end-over-end to measure the bridge's length, which is reported to be 364.4 Smoots plus one ear (Geeslin and Brown

1989, p. 93). Readers interested in a more complete discussion of units and measures are referred to Pennycuick (1988), Glover and Young (1996), Lide (2001), and Whitelaw (2007).

2.1 Angular Units

The following eight angular units are the ones most frequently encountered in surveying, mapping, geodesy, and geophysics.

2.1.1 Degrees, minutes, and seconds

Some historians believe that the modern division of the circle into 360 parts comes from the Sumerians by way of the Babylonians (Smith 1958; Maor 1998; Whitelaw 2007). The Sumerian number system appears to have evolved from two origins: one base 6 and the other base 10. This led to a base 60 number system. In this system, the first 60 numbers have their own symbols, as with the Arabic decimal system having the ten symbols 0 1 2 3 4 5 6 7 8 9. Apparently, the cycle of time served as a basis for the division of the circle into 360 parts because the year had about 360 days. This may have appealed to the ancients because subdividing 360 by 60 yields six equal parts, roughly two lunar months. Geometrically, a circle can be inscribed by a hexagon, a figure of six equilateral triangles each with three 60° angles (Maor 1998).

The 360 divisions of a circle are, of course, called **degrees**, from the Latin *de gradus* (Smith 1958). The origins of the word degree suggest a connection with a step on a ladder (Smith 1958; Maor 1998), which visually connects the image of the demarcations of a thermometer with the rungs of a ladder. Degrees are denoted by a small circle, e.g., 43°. This notation is a visual reminder that this quantity is 43/360 of a circle.

Keeping with dividing into sixty parts, the degree can be subdivided into 60 **minutes**, from the Latin *pars minuta prima*, or "first small part." Minutes of arc are denoted with a single tick mark, e.g., 43° 15′. Minutes are also subdivided into sixty parts, called **seconds**, from the Latin *pars minuta secunda*, or "second small part." Seconds of arc are denoted by two tick marks, e.g., 43° 15′ 52″ (read as "43 degrees, 15 minutes, 52 seconds"). Confusion can arise between the temporal units "minute" and "second" with the angular units so, when necessary, the angular units include an "arc" qualification (for example "minutes of arc" or "arc minutes").

Sometimes angles can be determined with precision better than integer arc seconds. For example, one second of latitude spans approximately 30 meters at sea level and GNSS positioning routinely achieves submeter precision, or even better. To express fractions of a second using this Babylonian-like system, we would subdivide seconds into a third part denoted by triple tick marks. However, this is not done. Instead, we meld the decimal system into the Babylonian system at any stage of the subdivision. Thus, an angle might be expressed in decimal degrees (DD); degrees and decimal minutes; or degrees, minutes, and decimal seconds (DMS). It is traditional to write decimal seconds with a double tick mark over a period. Computer fonts do not have this character, so decimal seconds are written sometimes as a decimal and sometimes with a double tick mark and a period (e.g., 18.″25).

Let δ (delta) denote the degrees of a DMS angle, μ (mu) denote the minutes, and σ (sigma) denote the seconds. To convert from a DMS angle $\delta° \mu' \sigma''$ to DD

$$DD = \delta + \frac{\mu}{60} + \frac{\sigma}{3600} \tag{2.1}$$

2.1. ANGULAR UNITS

For example, 73° 14′ 32″N in decimal degrees is

$$DD = 73 + \frac{14}{60} + \frac{32}{3600}$$
$$= 73 + 0.2333 + 0.0089$$
$$= 73.2422°$$

Converting from DD to DMS is slightly more complicated. It is easiest to explain it in terms of an algorithm[1] and an example. Let $DD = 317.215\,52°$. To convert DD to DMS

1. DMS degrees δ equals the integer part of the decimal degrees. Therefore, $\delta = 317°$.

2. DMS minutes μ equals the integer part of $(DD - \delta) \cdot 60$. Therefore, $\mu =$ the integer part of $(317.215\,52 - 317) \cdot 60 = 0.215\,52 \cdot 60 = 12.9312$, of which the integer part is 12. So, $\mu = 12$.

3. DMS seconds σ equals $(DD - \delta - \frac{\mu}{60}) \cdot 3600$. Therefore, $\sigma = (317.215\,52 - 317 - \frac{12}{60}) \cdot 3600 = 55\overset{''}{.}8720$.

Thus, $317.215\,52° = 317°\,12'\,55\overset{''}{.}8720$.

Converting negative DMS to decimal degrees is simply the negative of the positive conversion. For example, -73° 14′ 32″ in decimal degrees is

$$DD = -(73 + \frac{14}{60} + \frac{32}{3600})$$
$$= -(73 + 0.2333 + 0.0089)$$
$$= -73.2433°$$

Latitudes and longitudes have additional considerations concerning how their angles are denoted (see section 4.3 on page 58).

2.1.2 Radians

Radians are a natural angular unit because they arise from the relationship between a circle and its circumference. It is well known that the circumference of a circle is the circle's radius multiplied by twice π (Pi) or $2\pi r$. A circle whose radius is exactly one unit in length is called a **unit circle**. The circumference of a unit circle is 2π units, approximately 6.2832 units. As shown in Fig. 2.1, a **radian** is defined to be that angle such that the arc length of the circle segment subtended by this angle is exactly one unit in length. Thus, for a unit circle, an angle of one radian creates an arc of length 1. This gives rise to the relationship if r is the radius of a circle and $0 \leq \theta < 2\pi$ is some angle expressed in radians, then the length of the circle segment subtended by θ is

$$d = r\theta \quad (2.2)$$

Equation 2.2 is a fundamental concept in geometrical geodesy because it gives the relationship between distances on a sphere to angles. Rearranging Eq. 2.2 shows that the angle θ is given by a ratio of two lengths: the length of the circular segment and the radius of the circle. Consequently, radians can be thought of as having units of either length per length or, more commonly, no units

---

[1] An **algorithm** is a sequence of instructions that can be followed to accomplish some task.

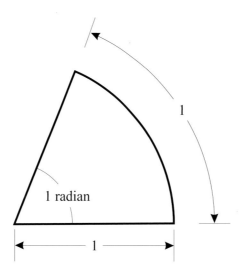

Figure 2.1: Definition of a radian.

at all. Also, the definition of the radian leads to the observation that 360° means the same thing as 2π radians. Therefore, to convert from radians to decimal degrees, simply multiply the angle in radians by $180°/\pi$. To convert from decimal degrees to radians, multiply the angle in decimal degrees by $\pi/180°$.

Example 2.1. Suppose we assume that the Earth is spherical and has a radius of 6 371 000 m, which is close to the average geometric distance from the Earth's center to sea level along a meridian from the equator to either pole. Determine the distance d along the surface of the spherical Earth from the equator to the North Pole.
Solution: The angle in radians from the equator to the North Pole is one quarter of a full circle, or $2\pi/4 = \pi/2$. So

$$\begin{aligned} d &= r\theta \\ &= 6\,371\,000\,\text{m} \times \pi/2 \\ &= 10\,007\,000\,\text{m} \end{aligned}$$

□

Example 2.2. Convert 97.802° to radians.
Solution: $97.802° \times \pi/180° = 1.707$ radians.
□

Example 2.3. Convert -0.1166 radians to DD.
Solution: $-0.1166 \times 180°/\pi = -6.6807°$.
□

Example 2.4. What is the arc length of the circular segment created by an angle subtending $147°19'23''$ in a circle whose radius is 198.26 meters?

2.1. ANGULAR UNITS 13

Solution: We must first convert the angle from DMS to DD and then to radians. The angle is given as $147°19'23''$, so this angle is

$$147 + 19/60 + 23/3600 = 147.323\,0556°$$
$$147.323\,055\,6° \times \pi/180° = 2.571\,272\,383\,55 \text{ radians}$$

The radius is given as 198.26 meters, so $d = r\theta = 198.26$ meters $\times 2.571\,272\,383\,55 = 509.78$ meters. □

2.1.3 Circles and semicircles

One **circle** equals 2π radians, or $360°$, and one **semicircle** equals π radians, or $180°$. Angles expressed in circles can have a very compact notation. If θ is an angle expressed in circles, then $0 \leq \theta < 1$. Therefore, all angles expressed in circles are decimals beginning with a zero followed by a decimal point so the zero and the decimal point can be suppressed. The fractional phase of an electromagnetic wave, a common GPS observable, is often expressed in circles. If an angle is known to not exceed $180°$, then semicircles provide twice the precision with the same compact notation. This is convenient for applications that need to express angles with a simple integer notation, such as the GPS navigation message.

2.1.4 Grade/grad/gon

The **grade**, also known as the **grad** or **gon**, is an angular unit dividing the circle into 400 parts, so 400 gon $= 2\pi$ radians. The grade originated in France as part of the movement to standardize all units to decimal divisions, so it assigns 100 divisions to one quarter circle, rather than 90 divisions as is done with degrees. One advantage of the gon is that it is easy to determine angles in excess of a right angle. For example, $35°$ more than $90°$ is $125°$ but 35 gon more than 100 gon is 135 gon. Conversely, 351 gon is 51 gon further rotation than 3/4 of a cycle. The digit in the hundreds place indicates which quarter circle an angle is in. A disadvantage of the gon is that certain common angles are not integers. For example, one-third and two-thirds of a right angle are exact integer numbers of degrees ($30°$ and $60°$, respectively) whereas they are $33.3\ldots$ and $66.6\ldots$ gons, respectively, which are nonterminating decimals.

Gons are often subdivided as centigons (100 centigrade = 100 centigon = 1 gon) and milligons (1000 milligrade = 1000 milligons = 1 gon). The obvious possible confusion between the angle "centigrade" and the temperature unit of the same spelling has led to grad and gon being far more common than grade. The international standard symbol for this unit is "gon," so one would write $90° = 100$ gon.

2.1.5 Mil

The **mil** is an angular unit developed by militaries to direct artillery fire. The NATO mil (including Canada) is 1/6400 circle. As such, one NATO mil equals $0.000\,981\,748$ radians, which is close to one milliradian, and it is from this relationship that the unit gets its name. There are 1600 NATO mil in one quarter circle. The army of the Soviet Union defined a mil to be 1/6300 circle, which subdivides $90°$ into 1575 mil. The army of the United Kingdom defined a mil to be 1/1000 radian, exactly one milliradian, so one quarter circle is 1570.8 U.K. mil. During World War II, the army of the United States defined a mil to be 1/4000 circle, so there are 1000 U.S. mil per quarter circle. The symbol of the mil is "mil."

2.1.6 Milli-arc second (mas)

A milli-arc second, abbreviated **mas**, is 1/1000 arc second, as the name implies. It is used in geodesy, for example, to describe the extremely small angular rotations needed to transform coordinates between different geodetic reference frames. The mas is also found in geophysics and astronomy literature. The symbol of the milli-arc second is "mas."

2.2 Linear Units

The radian and the circle can reasonably thought to be "natural" units for angles, but until recently, there was no obvious natural choice for a linear unit. There have been a large number of linear units defined by various odd things such as the length of body parts (foot, inch, yard), the size of the Earth, the length of camel strides, how far a person can throw an ax, and the distance someone can walk while smoking a bowl of Prince Albert tobacco in a pipe. Perhaps surprisingly, modern physics has allowed the definition of a very natural unit for length, but one that is actually derived from the measurement of time. The meter and the second share a definition that depends on something known to be constant, namely, the speed of light in a vacuum. Most of the material in this section comes from NIST (2003).

The speed of light in a vacuum is constant. This consistency has given rise to an astonishingly successful unit for length based on the following logic. Suppose it is possible to create a source of monochromatic light.[2] The velocity of a wave (c, length per time) is given by the product of its wavelength (λ, length per cycle) and the wave's frequency (f, cycles per time): $c = \lambda f$. So, if the velocity of a wave is constant and the wave is monochromatic, then the period of time between oscillations is also constant. Because the speed of light in a vacuum is a constant, being able to measure time extremely precisely leads to a very precise definition of distance in terms of how far a wave travels in a certain period of time. The definition of the meter, therefore, begins with the definition of the second.

2.2.1 Second

> The **second** (s) is the duration of 9 192 631 770 periods of the radiation corresponding to the transition between the two hyperfine levels of the ground state of the caesium-133 atom.

This definition is from the International Standards Organization (ISO) standard 31-1, which gives the name, symbol, and definition for 21 quantities and units of space and time. Although the definition of the second is highly technical, it simply means that a caesium-133 atom can be caused to vibrate and emit light (radiation) in such a way that those vibrations are exceedingly uniform in time. Consequently, the radiation emitted from two different caesium-133 atoms excited in this fashion will be nearly exactly equal in frequency, thus giving rise to a dependable, universal method of timing: counting the oscillations is tantamount to creating an exceedingly accurate timer. So, the second is then simply *defined* to be a certain number (9 192 631 770) of oscillations of such light (Kovalevsky et al. 1989, p. 9, 379–415).

---

[2] **Monochromatic** means single frequency. Lasers are the source of the most monochromatic light.

2.2.2 Meter

Perhaps the most common linear unit is the **meter** (m), being the internationally accepted standard. It is not an accident that the distance found in Example 2.1 is very nearly 10 000 000 m. At its inception in 1791, the meter was defined to be 1/10 000 000 the distance along a meridian of longitude through Paris from the equator to the North Pole (Alder 2002). In 1927 the meter was defined to be the length of a particular platinum-iridium bar kept in Paris. In 1960 the meter was defined by the wavelength of krypton-86 radiation and, in 1983, the current definition was adopted:

> The **meter** is the length of the path traveled by light in vacuum during a timer interval of 1/299 792 458 of a second.

Notice that this definition is actually a time measurement scaled by the velocity of light from which a spatial distance is inferred! It happens to correspond to approximately 30.663 318 99 periods of the aforementioned caesium-133 atom oscillations.

2.2.3 Foot

Civilian mapping in the United States is often done in linear units of either the **U.S. survey foot** or the **international foot**. According to the NIST, the U.S. Metric Law of 1866 gave the relationship 1 m = 39.37 inches. From 1893 until 1959, the yard was defined as being exactly equal to 3600/3937 m, and thus the foot was defined as being exactly equal to 1200/3937 m.

In 1959 the definition of the yard was changed to bring the U.S. yard and the yard used in other countries into agreement. Since then the yard has been defined as exactly equal to 0.9144 m, and thus the foot has been defined as exactly equal to 0.3048 m. At the same time it was decided that any data expressed in feet derived from geodetic surveys within the United States would continue to bear the relationship as defined in 1893, namely, 1 ft = 1200/3937 m. The name of this foot is U.S. survey foot (or just survey foot) whereas the name of the new foot defined in 1959 is international foot. The two are related to each other through the expression 1 international foot = 0.999 998 of a U.S. survey foot exactly.

2.3 Area

Areal units arise in geomatics for many purposes, such as parcel mapping, hydrological analyses, ecological studies, and demographics.

2.3.1 Chains and acres

Acres and hectares are two common units for area. A hectare is 10 000 square meters, so a square region that is 100 meters on a side has an area of one hectare: 100 m × 100 m = 10 000 m^2. Area has units of length squared. The definition of the hectare is completely in keeping with the "base everything on powers of ten" philosophy that guides the metric system in general, and the 100 × 100 definition makes the hectare easy to remember. In contrast, an acre is 4840 square yards or 43 560 square feet. Neither number is a perfect square, so some explanation is in order. First some relevant linear units.

The acre is a very ancient unit of area, coming from Sanskrit originally (Whitelaw 2007), meaning the size of an agricultural field. One definition of an acre is the amount of land a person could plow in one day with a team of oxen, with a workday apparently being about half the available amount of sunlight. Medieval plows were large, complicated machines that were pulled by teams of up to four pairs of oxen. Ox-pulled plows were difficult to turn, so it was best to plow in long straight furrows. The acre described fields that, in fact, tended to not be square. The ploughman would get the team going and then plow until the oxen needed to rest. This length became known as the **furlong**, being one furrow long. The ploughman used a rod to drive the team and that rod became a unit of measure called the **rod**, which was also called the **perche**[3] and the **pole**. The acre was a field one furlong in length and four rods in width. The ploughman would drive the team out-bound for one furlong, turn the plow, and drive the team back in time for lunch. Then he would repeat this once more in the afternoon, arriving back where he started at the end of the day. Consequently, medieval arable land tended to form long and narrow fields, which is why an acre is not a perfect square.

Although exact definitions do not exist, the length of the rod was somewhere between 16 1/2 and 25 feet, with 16 1/2 feet being common. A furlong was equivalent to 40 rods (660 feet). At 5280 feet, a mile is exactly 8 furlongs. So, if 1 acre = 1 furlong × 4 rods, then 1 acre = 40 rods × 4 rods. One acre is defined to be 43 560 square feet, which implies that one rod = 16 1/2 feet, being the modern commonly accepted value.

The **chain** was defined by the Rev. Edmund Gunter (1581–1626) to be four poles. So one furlong is 10 chains and an acre is a 1 chain × 10 chain area. This particular relationship makes measuring rectangular areas with chains very easy (Linklater 2002; McHenry 2008). In particular, the area of a rectangular region is the product of the average length of its opposite sides in chains divided by 10, almost (this isn't exactly true due to the Earth's curvature). For example, a 20 chain × 10 chain region is $20 \times 10 = 200/10 = 20$ acres. This simple relationship is the foundation of the U.S. Public Land Survey System in which vast areas of the western United States were subdivided into one square mile sections and then subdivided into quarter sections and those into quarter quarter sections, or 16 quarter quarter sections per section. So, a section was a composite of $4 \times 4 = 16$ quarter quarter sections. The sections were nominally 1 mile on a side = 8 furlongs = 80 chains divided by four quarter quarter sections = 20 chains on a side per quarter quarter section. Therefore, $20 \times 20/10 = 400/10 = 40$ acres. The chain is formally called "Gunter's chain," and real metal chains were created for measuring it. Being real metal chains, they were made up of a standardized series of 100 links, and the **link** became a linear unit defined as 1/100 of a chain, or 66/100 feet. Land is surveyed in survey feet in the United States, so a U.S. chain equals 66 survey feet.

2.4 Conversions

The following is a summary of exact relationships among units. First, some abbreviations:

- ft = foot, generically
- USIn = U.S. survey inch
- USFt = U.S. survey foot
- USYd = U.S. survey yard

---

[3]French for "pole."

2.4. CONVERSIONS

- IntIn = international inch

- IntFt = international foot

- IntYd = international yard

The following relationships are exact. They are given as ratios of integer values. It is a common practice to infer that a number that is written as a decimal is not known exactly whereas an integer or a ratio of integers is known exactly. Previous sections gave definitions for various units that do not follow this convention. Those definitions were copied verbatim from their sources, and will be recapitulated here for convenience.

$$USIn = \frac{100}{3937} \text{ m}$$
$$USFt = \frac{1200}{3937} \text{ m}$$
$$USYd = \frac{3600}{3937} \text{ m}$$

$$IntIn = \frac{254}{10\,000} \text{ m}$$
$$IntFt = \frac{3048}{10\,000} \text{ m}$$
$$IntYd = \frac{9144}{10\,000} \text{ m}$$

$$USIn = \frac{100}{3937}\frac{10\,000}{254} = \frac{1\,000\,000}{999\,998} \text{ IntIn}$$
$$USFt = \frac{1200}{3937}\frac{10\,000}{3048} = \frac{1\,000\,000}{999\,998} \text{ IntFt}$$
$$USYd = \frac{3600}{3937}\frac{10\,000}{9144} = \frac{1\,000\,000}{999\,998} \text{ IntYd}$$

1 rod = 1 pole = 1 perch = 16 1/2 ft (modern)
1 furlong = 40 rods = 660 ft
1 chain = 4 rods = 66 ft
1 acre = 10 chain × 1 chain = 10 chain^2
1 acre = 43 560 ft^2
1 statute mile (or just "mile") = 5280 ft
1 mile = 8 furlong
1 mile^2 = 640 acre = 80 chain × 80 chain
1 hectare = 100 m × 100 m = 10 000 m^2

2.5 Your Turn

Problem 2.1. Convert $35.48481°$ to DMS. Then, convert your answer back to DD to verify your answer.

Problem 2.2. Convert $29°\ 13'\ 52''$ to decimal degrees.

Problem 2.3. Convert $-45°\ 06'\ 21\overset{''}{.}6695$ to decimal degrees.

Problem 2.4. If a coordinate is $1\,265\,201$ survey feet, what is the coordinate in international feet?

Problem 2.5. The mil was defined to be helpful in directing artillery as the following example shows. By $d = r\theta$ we know that aiming an artillery piece θ radians in either direction will cause the shell to land a certain linear distance offset (d) in that direction because, if θ is small, then arc length is nearly equal to the length of the arc's chord (see Fig. 6.10 on page 120). If a howitzer is currently aimed at a place 1000 U.S. yards away and the aim is changed one U.S. mil to the right, how many yards to the right will the next round fall?

Problem 2.6. What is the length of a circular arc segment on a circle of 100 m radius subtended by a one arc second angle?

Chapter 3

Reductions and Computations for Plane Surveying

In plane surveying, maps are compiled from positions that are computed from angles and distances. The computations to determine positions involve **reductions**, being computations that change observed quantities to those needed to compile a map. Geometrical geodesy is, in many respects, a game of reductions and this chapter will introduce their purpose and application. It is the first step in the journey from a flatland to a round one.

This chapter also discusses the two fundamental problems in mapping: the **direct** and **inverse problems** in the context of plane surveying. An application of the direct problem converts reduced field observations into the spatial coordinates of the objects being mapped, which are then used to draw the map. The inverse problem is to derive distances and directions from geospatial coordinates.

3.1 Absolute versus Relative Positions

Location is not an absolute concept. Location can be described only with respect to some reference system (Kovalevsky et al. 1989, p. 1). A map can be drawn by picking any location on a sheet of paper as an origin and any direction as north, then picking a **cartographic scale** (see section 3.5.1 on page 28) and drawing features directly from angles and distances using rulers and protractors. The coordinates of the positions of such a map could also be computed and would reflect offsets from the chosen origin, so all quantities on the map are relative to that origin. The locations of the map features are known only within the context of the local origin, and everything on the map is relative to it. Therefore, such positions are **relative** positions.

Alternatively, surveyors can collect their observations using **survey markers** (Fig. 3.1), which provide coordinates of that location that are meaningful in a global sense, such as latitude and longitude or elevation. Markers that provide "horizontal coordinates" (latitude and longitude) are called **monuments**. Those that provide elevation are called **bench marks**. These markers also allow the survey to be oriented to north as defined by a meridian of longitude. Positions computed from observations controlled by such markers are **absolute** positions.

Absolute coordinate maps have the property that the same features shown on two different but overlapping maps will overlay properly on top of each other, within error tolerances, provided the coordinates are in the same datum and coordinate system. This property makes absolute coordinate maps ideal for use in a GIS because disparate data layers ought to be able to be "mixed

Figure 3.1: Brass cap bench mark survey marker set in a rock outcropping.

and matched" freely. Unfortunately, this is often not the case, and the typical reason is that the data were collected in different datums and projected into different coordinate systems. One of the goals of this book is to explain how to cope with this problem correctly. However, this chapter is concerned with relative coordinates. Absolute coordinates are the subject of chapter 5.

It may seem odd that maps are ever compiled from relative coordinates, because a common use for maps (such as road maps) is to find something whose location is unknown to us. Maps compiled with relative coordinates are generally no help with this task. However, relative-coordinate maps serve a valuable purpose. A topographic map of a construction site or a subdivision property boundary map (called a **plat**) are two examples of maps that are commonly compiled using relative coordinates. In many cases, such as construction design or home property boundaries, the location of the area shown on the map is already known by the people using the map. Like the blueprint of a house, such maps are used to describe the precise spatial relationships of interesting features on the ground, but the absolute location of any particular house built from the blueprint is assumed and, therefore, omitted. Furthermore, before GPS, it was more difficult to measure absolute positions than relative positions. A **total station** is one of the most common instruments used in map making, being a combination of the theodolite for measuring angles and an electromagnetic distance measurement (EDM) instrument. Total stations cannot determine absolute coordinates directly, so if a surveyor wanted to know the absolute coordinates of some site to be surveyed, the surveyor would need to find the nearest horizontal survey control marker and perform an operation called a traverse. The U.S. National Geodetic Survey (NGS) defines a **survey traverse** as

> a route and a sequence of observation stations between which distances and directions have been observed from field measurements and have been used in determining locations of the points.

In the case mentioned above, the route would be the path from the nearest monument to the remediation site. The sequence of points would be places where the surveyors set up the total

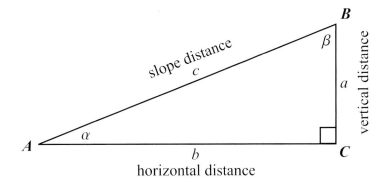

Figure 3.2: Slope, horizontal, and vertical distances. Vertices are indicated by uppercase Roman letters. The lengths of sides are named with lowercase Roman letters. Angles are indicated by Greek letters.

station to measure directions and distances. A traverse is the process by which coordinates are transferred from one location to another.

This chapter will focus on how to use trigonometry and relative coordinates to solve several kinds of mapping problems that arise when using relative positioning instruments such as total stations. These problems include computing the heights of objects without measuring them directly, computing angles and distances from coordinates, computing horizontal distances without measuring them directly, finding planimetric areas, computing the coordinates of a forward point, and staking out sampling transects. Such problems arise both in fieldwork and in the analyses performed with a GIS.

3.2 Linear Distances

The general term **distance** means the length of a curve segment. A **linear distance** is the length of a straight-line segment. Distances along circular arc segments arise frequently in geodesy and will be dealt with in chapter 6. There are three types of linear distance of interest in this chapter – horizontal, vertical, and slope distances (Fig. 3.2).

Many surveying instruments must be brought into level before being used; otherwise the observations taken using them will be incorrect. Instruments are leveled by adjusting the horizontal reference plane of the instrument with fine-adjust screws until the bubbles of spirit levels on the instrument are in the center of their vials. After this is accomplished, the horizontal reference plane is perpendicular to the local vertical, being the direction a plumb bob hangs; **vertical** is the direction the Earth's gravity field acts at a location (NGS 2009). A line is **horizontal** if it is perpendicular to vertical. At any location there is only one horizontal plane (assuming the direction gravity acts is constant) but there are infinitely many vertical planes: one for each direction.

A **horizontal distance** is a linear distance in the horizontal plane. A **vertical distance** is a linear distance in a vertical plane. **Slope distances**, also called **slant** and **spatial** distances, are linear distances in an arbitrary plane. The EDM in a total station measures slope distances by reflecting a laser beam off a target, often a corner reflector prism set atop a prism pole or a tripod (see Fig. 3.3 on the following page). The prism's mirrors are set nearly 90° to one another, forming a corner, or **trihedral**. Trihedrals always reflect light back in the same direction from which the

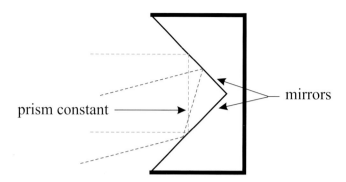

Figure 3.3: Light is always reflected by a trihedral prism parallel to the direction from which the light came.

light came (see the dotted lines in the figure). This is very handy because, otherwise, it would be necessary to aim the prism towards the EDM very precisely, which is extremely inconvenient. The trihedral shape adds a short leg to the path the light travels, adding an additional distance called the **prism constant**, which is shown as the line segment connecting the incoming and outgoing light paths. The prism constant must be subtracted from the distance reported by the EDM, which is usually done by the EDM itself. Some EDMs can measure distances without a prism target and, thus, are called "reflectorless EDMs."

3.3 Plane Angles

A plane angle is the rotational offset between two intersecting lines. There are many types of angles, as will be discussed below.

3.3.1 Pointings, directions, and angles

Theodolites are used to measure horizontal and vertical angles. Some angle measuring instruments contain graduated circles that are read by the operator and others by the instrument itself. **Pointing** the telescope at some aim point and reading the circle results in a **direction**, which is read in some angular unit, such as degrees, minutes, and seconds of arc. Directions usually fall in the range from $0°$ to $360°$. Two pointings result in two directions, and the difference of two directions is an **angle**. For example, if the circle for the first pointing reads $101°$ and the circle for the second pointing reads $154°$, then the angle between the two aim points is $154° - 101° = 53°$. It should be emphasized that, as just described, angles are computed from observed directions, but hereafter we will use the term "observed angle" as a shorthand for the difference of two directions.

3.3.2 Vertical angles

Vertical angles are formed by the intersection of two lines in a vertical plane (Fig. 3.4). Vertical angles are categorized as **zenith angles** and **angles of elevation** (AOE). We will use ζ (zeta) to denote zenith angles. Zenith angles use the vertical direction as a datum, so the up direction is a zenith angle of $0°$ and the horizontal direction is a zenith angle of $90°$. The situation is reversed for angles of elevation: the horizontal direction is an AOE of $0°$ and the vertical direction is an AOE of

3.3. PLANE ANGLES

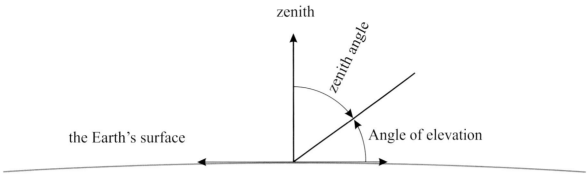

Figure 3.4: Zenith angles and angles of elevation.

90°. Zenith angles have a domain from 0 to π radians (0° to 180°). This is written mathematically as $\zeta \in [0, \pi]$. In this notation, the square bracket denotes a closed interval, and a parenthesis would denote an open interval, so $\zeta \in [0, \pi]$ is equivalent to $0 \leq \zeta \leq \pi$. Angles of elevation are in the range $[-\pi/2, \pi/2]$.

3.3.3 Horizontal angles

Horizontal angles are formed by the intersection of two lines in a horizontal plane. Horizontal angles are categorized as bearings and azimuths. According to the Geodetic Glossary (NGS 2009), in general a **bearing** is "the horizontal angle between a line from the observer to a given point, and a line from the observer along a specified direction (such as north)." I will use β (beta) to denote bearings. Bearings are usually taken to be in the range $\beta \in (-\pi, \pi]$. An **azimuth** is a bearing specified from a meridian and reckoned positive clockwise, with either south or north being used as zero. It is more common to use north as zero, so this convention will be followed throughout this book. I will use α (alpha) to denote azimuths. Azimuths are usually reported in the range $[0, 2\pi)$ but can take values of the entire real line in computations. This will happen, for example, if a total station is pointed at an azimuth of 350° and it is rotated 30° to the operator's right, resulting in a new pointing of 380°. Trigonometric functions are implemented in computer programs to allow input angles that exceed $\pm 2\pi$, so azimuths that exceed $\pm 2\pi$ are usually not problematic in computations, but they are never reported that way.

3.3.4 Spatial angles

Horizontal angles are defined in the horizontal plane, and vertical angles are defined in a vertical plane. Angles in an arbitrary plane are called **spatial angles**. Spatial angles often arise in a global context, such as the angle between two GPS satellites with respect to the center of the Earth.

3.4 Mathematical Tools

3.4.1 Sine, cosine, and tangent

This chapter is concerned with plane surveying and, therefore, the mathematics needed come from plane trigonometry, so it is appropriate to begin with a review of the basics. **Right triangles** are triangles that contain one 90° angle. The side opposite the right angle is called the **hypotenuse**, and it is always the longest side of the triangle. In Fig. 3.2, the right angle is the angle at C and the slope distance is the hypotenuse.

The fundamental trigonometric functions are the sine, cosine, and tangent, abbreviated sin, cos, and tan, respectively. For a right triangle, the sine of an angle is the length of the side opposite to it divided by the length of the hypotenuse. For the angle α in Fig. 3.2 $\sin \alpha = a/c$, and $\sin \beta = b/c$. The cosine of an angle is the length of the side adjacent to it divided by the length of the hypotenuse. Thus, $\cos \alpha = b/c$, and $\cos \beta = a/c$. The tangent of an angle is the length of the side opposite divided by the length of the side adjacent. So $\tan \alpha = a/b$, and $\tan \beta = b/a$. The reciprocal of the cosine is called the secant, the reciprocal of the sine is called the cosecant, and the reciprocal of the tangent is called the cotangent. These are abbreviated sec, csc, and cot, respectively.

3.4.2 Law of sines

The definitions for the sine, cosine, and tangent apply only to right triangles. However, right triangles seldom arise in nature, so there is a need to cope with triangles possessing no right angles. Two main formulæ are useful in this situation: the law of cosines and the law of sines. There are circumstances in which two angles and one distance are known and, in that case, we use the law of sines to determine the unknown quantities. The law of sines states that the ratio of the sine of an angle with the length of the side opposite it is a single constant for all angles of triangle, which can be expressed mathematically as

$$\frac{\sin \theta_A}{a} = \frac{\sin \theta_B}{b} = \frac{\sin \theta_C}{c} \qquad (3.1)$$

Example 3.1. Referring to Fig. 3.5 on the next page, suppose $a = 39$ m, $\theta_A = 32°$ and $\theta_B = 58°$. Find b.

Solution: From Eq. 3.1

$$\frac{\sin \theta_A}{a} = \frac{\sin \theta_B}{b}$$
$$\frac{\sin 32°}{39} = \frac{\sin 58°}{b}$$
$$b = \frac{39 \sin 58°}{\sin 32°}$$
$$b = 62.413$$

Side b has units of meters because a was given in meters and the trigonometric functions are unitless.
□

3.4. MATHEMATICAL TOOLS

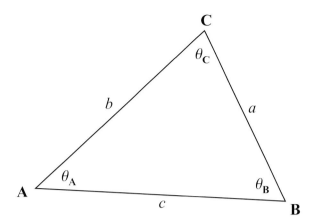

Figure 3.5: How the symbols relate to the parts of a triangle in the law of cosines.

3.4.3 Law of cosines

The law of cosines states that the cosine of any angle of a triangle is the following ratio of the lengths of the sides of a triangle:

$$\cos\theta_C = \frac{a^2 + b^2 - c^2}{2ab} \qquad (3.2)$$

Please refer to Fig. 3.5. Total stations measure angles (such as θ_A, θ_B, and θ_C) and distances (such as a, b, and c). Thus, if we need to know the distance between two points **A** and **B** and we cannot measure it directly, we use the following method:

1. Set up a total station on some point **C** such that both **A** and **B** are visible.

2. Determine the horizontal distance from **C** to **A**. This is the distance b (recall that b is the length of the triangle's side opposite **B**).

3. Determine the horizontal the distance from **C** to **B**. This is the distance a (recall that a is the length of the triangle's side opposite **A**).

4. Determine the angle from **A** to **B**. This is the angle θ_C.

5. Use some algebra to rearrange Eq. 3.2 and solve for c:

$$c = \sqrt{a^2 + b^2 - 2ab\cos\theta_C} \qquad (3.3)$$

6. Plug the given values into Eq. 3.3 to determine the desired length, c.

Example 3.2. Referring to Fig. 3.5, suppose $c = 301$ m, $b = 285$ m, and $\theta_A = 32°$. Find a.
Solution: The variables in Eq. 3.3 can be rearranged to solve for the length of any side, so

$$\begin{aligned} a &= \sqrt{b^2 + c^2 - 2cb\cos\theta_A} \\ &= \sqrt{285^2 + 301^2 - 2(285)(301)\cos 32°} \\ &= \sqrt{81\,225 + 90\,601 - 171\,570 \times 0.848\,048} \\ &= 162 \end{aligned}$$

□

3.4.4 Computing area

In some cases, area needs to be computed without considering any undulations or slope of the terrain, such as the area of a parcel of land for taxation purposes. In other cases, the three–dimensional nature of the terrain is of interest, such as for hydrological studies. If the three–dimensional ground has been depicted using horizontal angles and distances, then it has been projected into a plane and is known as a **planimetric** representation. Maps are compiled on topologically planar surfaces, therefore grid coordinates and other features in maps are said to be planimetric, i.e., referred to a plane. Planimetric area is always less than or equal to slope area because horizontal distances are always less than or equal to slope distances. This fact is known as the **triangle inequality**.

Planimetric area

Suppose the planimetric area of a dairy pasture is needed to lay out a vegetation sampling transect. The pasture has an irregular shape (see Fig. 3.6 on the next page). We will approach this problem by subdividing the region of interest into a tessellation[1] of triangles, computing the area of each triangle, and summing them to determine the area of the whole.

As shown in Fig. 3.6, we can compute the area of the pasture as the sum of the three subareas. So, it remains to see how to compute the area of an oblique triangle. The formula for the area of a right triangle ($\frac{1}{2} \times$ base $\times$ height) is of no use because none of these triangles are right triangles. Consider a planar triangle whose sides have lengths a, b, and c. Define a constant s as

$$s = \frac{1}{2}(a+b+c) \tag{3.4}$$

Then the area A of the triangle is

$$A = \sqrt{s(s-a)(s-b)(s-c)} \tag{3.5}$$

Example 3.3. Determine the area of a triangle with the following sides: $a = 912.71$ m, $b = 583.58$ m, and $c = 763.14$ m.
Solution: From Eq. 3.4, $s = \frac{1}{2}(912.71 + 583.58 + 763.14) = 1129.715$ m. From Eq. 3.5 we have

$$\begin{aligned} A &= \sqrt{s(s-a)(s-b)(s-c)} \\ &= \sqrt{1129.715(1129.715 - 912.71)(1129.715 - 583.58)(1129.715 - 763.14)} \\ &= 221\,539.28 \text{ m}^2 \end{aligned}$$

□

Slope area

In many applications, such as hydrology, we are interested in the actual three–dimensional area of the ground, not just the projected area presented in the previous section. It happens that it is just as simple to compute slope areas as planimetric areas. The only adjustment to be made is in the lengths of the sides of the triangles: instead of using horizontal distances, use slope distances. Everything else is the same.

---

[1] A **tessellation** is a set of polygons that have no gaps between their edges and completely cover the area of interest, like tiles on a floor.

3.4. MATHEMATICAL TOOLS 27

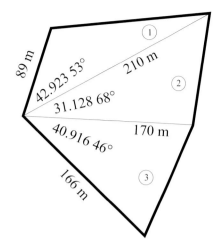

Figure 3.6: Pasture tessellated into three triangular subregions.

Example 3.4. Suppose we measure the slope distances along the three sides of a triangle from the previous example to be $a = 1005.91$ m, $b = 589.66$ m, and computed $c = 823.90$ m. Find the slope area of this triangle.
Solution: From Eq. 3.4, $s = \frac{1}{2}(1005.91 + 589.66 + 823.90) = 1209.74$ m. From Eq. 3.5 we have

$$\begin{aligned} A &= \sqrt{s(s-a)(s-b)(s-c)} \\ &= \sqrt{2419.47(2419.47 - 1005.91)(2419.47 - 589.66)(2419.47 - 823.90)} \\ &= 242\,883 \text{ m}^2 \end{aligned}$$
□

In this case the slope area is approximately 1.1 times larger than the planimetric area.

3.4.5 Inverse trigonometric functions

Suppose the goal is to compute an angle from the lengths of the sides of a triangle. For this problem, we use the inverse (or "arc") trigonometric functions: the arcsine, arccosine, and arctangent, which are sometimes written as a reciprocal, such as $\sin^{-1}\theta$. This might be a confusing convention because the inverse trig functions are not reciprocals: $\sin\theta \times \sin^{-1}\theta \neq 1$. Rather, the inverse trig functions return the angle when given the ratio, whereas the trigonometric functions return the ratio when given the angle. Expressed mathematically, $\sin^{-1}(\sin\theta) = \theta$.

Example 3.5. Suppose a right triangle has a side a of length 10 and a hypotenuse of length 20. What is the angle opposite a?
Solution: $\sin\theta = 10/20 = 0.5$ and $\sin^{-1}(\sin\theta) = \theta$, so $\theta = \sin^{-1}(\sin\theta) = \sin^{-1}(0.5) = 30°$.
□

3.4.6 Pythagoras's theorem

Pythagoras's famous theorem establishes the relationship among the lengths of the sides of right triangles: the square of the length of the hypotenuse equals the sum of the squares of the other two

sides. In equation form,
$$c^2 = a^2 + b^2$$
where c is the length of the hypotenuse and a and b are the lengths of the other two sides. This equation leads to a fundamental geometric relationship called the **triangle inequality**: the length of the hypotenuse is less than the sum of the other two sides. The next section uses the Pythagorean theorem to solve one of the most common map use problems, the inverse problem in the plane.

3.5 The Inverse Problem in the Plane

One often wants to determine the distance and direction between two locations: how far is London from Sydney, and in what direction does one go to get there? Mappers call this problem **inversing**, and it is the opposite problem to the one they confront most of the time, which is to say, given the coordinates of a known point and the distance and bearing to an unknown point, find the unknown point's coordinates. The latter is known as the **direct** (also known as the **forward**) problem and will be covered in section 3.7. GIS practitioners are inversing, perhaps unknowingly, when they use an interactive measuring tool on the computer screen. Such a tool is typically used by clicking on some feature of interest, then dragging the tool to some other feature of interest with a display of distance and azimuth changing as the tool moves. It happens that many common uses of a map involve inversing, so this section will review basic problems that arise when using a map, illustrating how inversing solves those problems.

3.5.1 Map scale

Map scale is a ratio whose denominator is some distance on the ground and whose numerator is an equivalent distance on the map. This ratio is called the **representative fraction** (RF) and also the **map scale ratio** (MSR). For example, if one inch on the map is equivalent to 24 000 inches on the ground, then the MSR is 1/24 000, which is usually written 1:24 000. The units of the MSR are somewhat subtle: it is a distance divided by a distance, so some authors write that MSR is unitless. However, these are different *types* of distances, namely map linear units divided by ground linear units. For example, an MSR can reasonably be said to have units of "inches on the map per inch on the ground." This example used inches but could have used meters, furlongs, chains, or any other linear unit because the ratio is generic so long as the map units are the same as the ground units. For example, for a 1:24 000 scale map: 1 inch on the map corresponds to 24 000 inches on the ground, and 1 centimeter on the map equals 24 000 centimeters on the ground. However, it is possible the linear unit measured on the map is not the same as the linear unit on the ground. Although measuring distances on the map in inches, for example, is convenient, inches are unlikely to be convenient units on the ground – feet would be more likely. We are at liberty to state the MSR numerator and denominator in different units. The map could be compiled at 1 map inch = 24 000 ground inches and express the ground inches in ground feet by dividing by 12 inches/foot giving 1 map inch = 2000 ground feet. This would be a 1 inch : 2000 feet map scale and is completely equivalent to 1 map inch = 24 000 ground inches. Scales with different linear units in the MSR numerator and denominator are very common in maps used for construction, which are often compiled with scales such as 1 inch = 20 feet. The possibility of having disparate units in the MSR numerator and denominator illustrates the importance of having a clear understanding of the MSR's units.

3.5. THE INVERSE PROBLEM IN THE PLANE

Maps are referred to as being large-scale or small-scale, and these terms can be counterintuitive. A large-scale map is one whose MSR is a large number; conversely a small-scale map is one whose MSR is a small number. Confusion arises because large-scale maps cover small amounts of area compared to small-scale maps. The reason is that an MSR is a fraction. Consider an MSR of 1/24 000 compared to 1/1 000 000. The former is a much larger number than the latter, so a 1:24 000 map is at a much larger scale than a 1:1 000 000 map. There can be a tendency to look at the magnitude of the denominator of the MSR and think that 1 000 000 is much larger than 24 000 so 1:1 000 000 must be a large scale. But the MSR is the *inverse* of this number, so the larger the denominator, the smaller the scale. It can also be helpful to realize that a (relatively) large-scale map requires a larger sheet of paper to depict the same area as a smaller-scale map.

Scaling

To determine the distance between two features shown on a map, measure the distance between them using a ruler. Let that distance be denoted by d_m, for "distance on the map." Convert d_m to ground units by dividing it by the MSR. This process is called **scaling** coordinates.

Example 3.6. We measure two features on a 1:5000 map and find them to be separated by 0.048 m. How far apart, in meters, are the features in the real world?
Solution: The MSR is 1 map meter/5000 ground meters. Dividing 0.048 map m by (1 map meter/5000 ground meters) is the same as multiplying 0.048 map meters by 5000 ground meters/map meter, yielding 240 ground meters. □

Example 3.7. A map is published with an MSR of 1:24 000. Suppose two features are separated by 3.41 inches. How far apart are these features on the ground in feet?
Solution: 3.41 map inch /(1 map inch/24 000 ground inch) = 3.41 map inch × 24 000 ground inch / 1 map inch = 81 840 ground inches. Converting to feet: 81 840 ground inches/(12 inch/foot) = 6820 ground feet. □

Instead of dividing map distances by the MSR to obtain ground distances, it can be easier to multiply by the denominator of the MSR when the numerator is one.

Example 3.8. Suppose a 1:50 000 map is measured with a scale ruled in millimeters. What is the shortest ground distance in meters that can be resolved?
Solution: Converting to meters, 1 map mm = 0.001 map meter. 0.001 map meter × 50 000 ground meters/1 map meter = 50 ground meters. Therefore, it would not be sensible to report distances scaled with this ruler at finer than a 50-meter resolution. □

This example illustrates a very important principle in mapping: in general, do not report spatial quantities with more significant digits than is warranted given the precision at which they were determined. The finest marked division is called the **least count** of the instrument. If the vertical circle in a total station has a least count of 5 inches, then it is good practice to report the angles observed with this instrument rounded to the nearest 5 inches, as shown in Fig. 3.11 on page 41.

Example 3.9. Suppose a 1:24 000 map is measured with a scale whose finest ruling is 1/32 of an inch. Two features are found to be 6 17/32 inches apart on the map. How far apart are they on the ground, in feet?
Solution: 6 17/32 map inches × 2000 ground feet/1 map inch = 13 062.5 ground feet. However, 1/32 inch × 2000 ground feet/1 map inch = 62.5 feet, so the shortest distance on the ground that

Figure 3.7: A scale bar shown in the margin of a map. This graphic indicates the distance on the map that represents up to 500 meters on the ground in 100-meter intervals.

can be resolved with a 1/32-inch rule is 62.5 feet. The distance separating the features is not known with greater certainty than this, so it should be reported rounded to the nearest 100 feet to be conservative, yielding 13 100 feet as the answer. □

One is sometimes confronted with using a map of unknown scale. This happens, for example, with enlarged or reduced photocopies of the original map, and with paper maps shrunk by humidity. Recall that MSR is map distance/ground distance. If there is something on the map with a known length or there are two point features separated by a known distance, then we can measure the separation on the map and divide it by the distance on the ground. A scale bar is perfect for this (Fig. 3.7).

Example 3.10. Suppose a map's scale bar shows 500 m, which is measured with a ruler to be 5.28 inches long. What is the map's scale?
Solution: Convert map distance to the same units as ground distance: 5.28 inch/12 (inch/ft) × 1200/3937 (ft/inch) = 0.134 112 m. So, the MSR is 0.134 112 m/500 m = 0.000 268 224. Although this is the answer, this number does not resemble the other MSRs shown so far, such as 1/24 000, because it's written as a decimal fraction rather than a ratio. To write a decimal fraction as a ratio, compute its reciprocal and use that as the denominator of the MSR. In this case, 1/0.000 268 224 = 3728, so the map's scale is 1:3728.
□

Determining scale is a common task in **photogrammetry** (Wolf and Dewitt 2000), which is defined by NGS as "The science of deducing the physical dimensions of objects from measurements on images (usually photographs) of the objects." Aerial photographs have no annotations, such as scale bars, that indicate scale. An image might contain a feature of known (or measurable) length, such as a runway. However, many aerial photographs have no such features, such as a photograph of an undeveloped prairie. Highly visible panels are often placed on the ground ahead of the photogrammetry flight so that they will appear in the photographs. The panels' locations are surveyed so that their centers have well-defined positions. The distance between the panels can be determined from the positions' coordinates by computing the planimetric distance between them.

3.5.2 Planimetric distance

To find the planimetric distance d_h between two locations **A** and **B** whose coordinates are $(e_\mathbf{A}, n_\mathbf{A})$ and $(e_\mathbf{B}, n_\mathbf{B})$, respectively, use the Pythagorean theorem:

$$d_h = \sqrt{(e_\mathbf{A} - e_\mathbf{B})^2 + (n_\mathbf{A} - n_\mathbf{B})^2} \tag{3.6}$$

Equation 3.6 holds for both relative and absolute planimetric coordinates (but not geographic coordinates). The reason that the Pythagorean theorem is applicable is because, in planimetric

3.5. THE INVERSE PROBLEM IN THE PLANE

(i.e., Cartesian) coordinate systems, the axes are mutually perpendicular. Therefore, the change in eastings is a distance in a direction that is necessarily perpendicular to the change in northings, thus forming a right triangle, as is required for the Pythagorean theorem to hold.

Example 3.11. An aerial photograph has ground control panels whose centers are **A** and **B** with coordinates given in meters as $(261\,201.54, 346\,515.02)$ and $(261\,520.60, 345\,584.53)$, respectively. Find the planimetric distance between **A** and **B**.

Solution: From Eq. 3.6

$$\begin{aligned}
d_h &= \sqrt{(e_{\mathbf{A}} - e_{\mathbf{B}})^2 + (n_{\mathbf{A}} - n_{\mathbf{B}})^2} \\
&= \sqrt{(261\,201.54\,\mathrm{m} - 261\,520.60\,\mathrm{m})^2 + (346\,515.02\,\mathrm{m} - 345\,584.53\,\mathrm{m})^2} \\
&= \sqrt{(-319.06\,\mathrm{m})^2 + (930.49\,\mathrm{m})^2} \\
&= \sqrt{101\,799\,\mathrm{m}^2 + 865\,812\,\mathrm{m}^2} \\
&= 983.67\,\mathrm{m}
\end{aligned}$$

□

Notice that the control coordinates are reported to two significant digits after the decimal, so the distance between them is reported to two significant digits, too.

Scaling map grid coordinates

Many maps are compiled showing coordinate system grids that aid the map reader in locating features. U.S. Geological Survey (USGS) 7.5-minute topographic maps (informally called "quad sheets") have latitude and longitude coordinates written next to tick marks along their borders. In fact, a quad sheet's borders (called the **neat lines**) depict lines of constant latitude and longitude, both being separated by 7.5 minutes of arc. The geographic coordinate grid is implied by the tick marks; no grid lines, apart from the neat lines, are drawn on the map. However, many quad sheets are compiled showing a map projection grid whose lines are spaced apart by some convenient distance, such as one kilometer for the Universal Transverse Mercator (UTM) grid.

Suppose someone wanted to know the grid coordinates of some map feature, perhaps to input as a waypoint in a GPS receiver. Let's assume that the map has a grid indicating lines of constant eastings and northings in the UTM grid coordinate system. For now, think of an easting as being a Cartesian coordinate oriented positive in the east direction, and of a northing as being a Cartesian coordinate oriented positive in the north direction (see section 8.4 on page 160 for more information about grid coordinates). To determine the feature's grid coordinates, one could locate the grid intersection that is south and west of the feature of interest, read that intersection's coordinates off the map, and then use a drafting triangle with a 90° angle to mark lines parallel to the grid lines through the point of interest. The parallel lines are offset from the southwest corner by two distances; call them δe_m for the easting offset and δn_m for the northing offset.[2] The corresponding offsets' lengths on the ground, δe_g and δn_g, are found by measuring δe_m and δn_m with a scale, dividing them by the MSR to produce two ground distances, and performing a units conversion to

---
[2]δ is the the lowercase Greek letter *delta*. δ usually denotes some discrete offset of any size whereas Δ, the capital Greek letter *delta*, usually denotes a very small discrete offset.

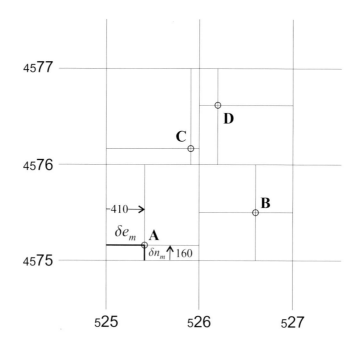

Figure 3.8: Grid lines and features on a map compiled with a UTM grid.

be compatible with the grid units, if necessary (meters for UTM). The feature's grid coordinates are computed by adding δe_g and δn_g to the grid intersection's easting and northing coordinates, respectively. This process is called **scaling** coordinates.

Example 3.12. Use scaling to determine the UTM grid coordinates of feature **A** in Fig. 3.8. Notice how the grid coordinates' leading digits are written in a smaller font in the figure. This is because those digits do not change across the map; the thousands digits, which do change, are emphasized. Notice also that the three trailing zeros are suppressed: these are never nonzero digits because the grid lines are spaced 1000 m apart. So 4 575 000 is written 4575, and 525 000 is written 525, etc.

Solution: The closest horizontal grid line south of **A** has a northing of 4 575 000 m. δn_m is measured, and after dividing by the MSR and applying a unit conversion, gives $\delta n_g = 160$ m. Since this grid line is south of **A**, adding 160 m to 4 575 000 m makes **A**'s northing coordinate 4 575 160 m. Next, the closest vertical grid line west of **A** has an easting of 525 000 m. δe_m is measured, and dividing by the MSR and applying a unit conversion, gives $\delta e_g = 410$ m. Since this grid line was west of **A**, adding 410 m to 525 000 m makes **A**'s easting coordinate 525 410 m.
□

3.5.3 Map north and geodetic north

Geodetic north is the direction along a meridian towards the North Pole, whereas **grid north** is the direction of the vertical grid lines on a map. As will be discussed in chapter 5, grid north is usually not geodetic north, the angular difference being known as **grid declination** or **convergence**. Therefore, azimuths deduced from a map are generally not geodetic azimuths; they are **grid azimuths**.

3.5. THE INVERSE PROBLEM IN THE PLANE

Table 3.1: Angular adjustment κ in Eq. 3.8.

δe	δn	Quadrant Adjustment
+	+	0
+	−	180°
−	+	360°
−	−	180°

Computing grid azimuths

Grid azimuths can be measured directly from a map using a protractor, or computed between any two points **A** and **B** whose coordinates are given in a planimetric coordinate system. If $\delta e = e_\mathbf{B} - e_\mathbf{A}$ and $\delta n = n_\mathbf{B} - n_\mathbf{A}$, the grid azimuth $\alpha_{\mathbf{A},\mathbf{B}}$ from point **A** to point **B** is given by

$$\alpha_{\mathbf{A},\mathbf{B}} = \arctan[\delta e/\delta n] \tag{3.7}$$

Careful attention must be paid to the order of the operands of Eq. 3.7 because $\alpha_{\mathbf{A},\mathbf{B}} = 180° + \alpha_{\mathbf{B},\mathbf{A}}$. Also, it is clear from its definition that an azimuth can be any angle from 0° to 360°. Therefore, we have to make sure that we adjust the result of Eq. 3.7 to place it in the proper quadrant. Most calculators have an arctangent function that returns an angle in the range from −90° to 90°. This is inadequate, so an adjustment is necessary. Table 3.1 shows the amount to add to the result of Eq. 3.7 based on the signs of δe and δn. For example, if both δe and δn are negative, add 180° to the result of Eq. 3.7. To emphasize this adjustment and to account for when $\delta n = 0$, we write Eq. 3.7 as

$$\alpha_{\mathbf{A},\mathbf{B}} = \begin{cases} 90° & \text{if } \delta n = 0 \text{ and } \delta e > 0 \\ 270° & \text{if } \delta n = 0 \text{ and } \delta e < 0 \\ \arctan[\delta e/\delta n] + \kappa & \text{if } \delta n \neq 0 \end{cases} \tag{3.8}$$

where κ is the adjustment.

Example 3.13. Find the grid azimuth from **A** to **B** whose coordinates are given in Example 3.11 on page 31.
Solution: From Eq. 3.8

$$\begin{aligned}
\alpha_{\mathbf{A},\mathbf{B}} &= \arctan[(e_\mathbf{B} - e_\mathbf{A})/(n_\mathbf{B} - n_\mathbf{A})] + \kappa \\
&= \arctan[(261\,520.60 - 261\,201.54)/(345\,584.53 - 346\,515.02)] + \kappa \\
&= \arctan[319.06/(-930.49)] + \kappa \\
&= -18.9266° + \kappa
\end{aligned}$$

From Table 3.1, because $\delta e > 0$ and $\delta n < 0$, we must add $\kappa = 180°$ to the result: $-18.9266° + 180° = 161.073°$.
□

Example 3.14. Find the grid azimuth from **B** to **A** whose coordinates are as given in Example 3.11.

Solution: From Eq. 3.8

$$\begin{aligned}\alpha_{B,A} &= \arctan[(e_A - e_B)/(n_A - n_B)] + \kappa \\ &= \arctan[(261\,201.54 - 261\,520.60)/(346\,515.02 - 345\,584.53)] + \kappa \\ &= \arctan[(-319.06)/930.49] + \kappa \\ &= -18.9266° + \kappa\end{aligned}$$

Notice that the result of the arctangent so far in this example exactly equals that from the previous example. This is true because the arguments of the arctangent function in the two examples are, in fact, equal. This illustrates the quadrant ambiguity that we must resolve manually. From Table 3.1, because $\delta e < 0$ and $\delta n > 0$, we must add $\kappa = 360°$ to the result: $-18.9266° + 360° = 341.073°$. Note that adding $180°$ to $161.073°$, the result from Example 3.13, gives $341.073°$.
□

3.5.4 Converting bearings to azimuths

It is often the case that the ultimate goal of measuring bearings and distances is to produce the coordinates of the features that were measured. These coordinates can then be put into a GIS and mapped. This process depends upon being able to convert the bearings that were measured with a theodolite into azimuths, either in a local or in an absolute coordinate system. In a relative coordinate survey, some arbitrary direction will be defined to be local north. This direction and the coordinates assigned to the **point of beginning** (POB), being the first station occupied in a survey, define a Cartesian coordinate system.

The following method can be used to convert a bearing into an azimuth. Suppose there is a total station at marker **U**; a prism at marker **T**, called the **backsight**; and another prism at marker **V**, called the **foresight**. The backsight is so named because, over the course of a traverse, the surveyor will look backwards to that mark, having already visited it. In this case, at the beginning of a survey, the backsight is a mark used to orient the survey and is often not occupied – it is often a highly visible, distant object such as a radio tower; whereas the foresight is usually a mark to be occupied. Let $\beta_{T,U,V}$ denote the bearing from the backsight to the foresight and let $\alpha_{U,T}$ denote the azimuth from the total station to the backsight. Then, the azimuth from the total station to the foresight $\alpha_{U,V}$ is $\alpha_{U,T} + \beta_{T,U,V}$. The result of this sum can exceed $360°$. If it does, then $360°$ should be subtracted from it to return it to the range $0 \leq \alpha_{U,V} < 360°$.

Example 3.15. Let $\alpha_{U,T} = 37°$ and $\beta_{T,U,V} = 212°$. Then, $\alpha_{U,V} = 37° + 212° = 249°$. □

Example 3.16. Let $\alpha_{U,T} = 343°$ and $\beta_{T,U,V} = 79°$. Then, $\alpha_{U,V} = 343° + 79° - 360° = 62°$. □

3.6 Reductions for Plane Surveying

3.6.1 Reducing slope distances to horizontal distances

Consider the distance from a ground point **A** to a point **B**, which is atop a nearby building (see Fig. 3.2 on page 21). If we construct a right triangle with a hypotenuse from **A** to **B**, with a third point **C** directly below **B** and horizontal to **A**, then the distance $\overline{AC}$ is the horizontal distance and the distance $\overline{BC}$ is the vertical distance. Horizontal distance can be determined from an observed slope distance and vertical angle by reducing the slope distance into the horizontal plane. This

3.6. REDUCTIONS FOR PLANE SURVEYING

process is a "reduction" because slope distances are generally longer than horizontal distances. As a formula we have

$$d_h = d_s \sin \zeta \tag{3.9}$$

where d_h is a horizontal distance, d_s is a slope distance, and ζ is a zenith angle.

Height of instrument

The following examples illustrate the sort of computations done with total station observations. For a total station, the **height of the instrument** (abbreviated "hi") is the vertical distance from a point over which the instrument has been set up to the optical center of its telescope. For a prism, the **height of the rod** (hr) is the vertical distance from a point over which the prism has been set up to the optical center of the prism. Height of the rod can be a bit of a misnomer because it includes the height of the prism; it's not just the prism pole. The hr term is generic because it also includes prisms set on any support mechanism, like a tripod, and not just on poles. The abbreviations hi and hr are lowercase deliberately because the uppercase equivalents mean something different. HI is the height of an instrument above a datum, such as mean sea level; likewise for HR. For example, if a total station were set up over a bench mark whose elevation (height above mean sea level) was 1520 m and the hi was 1.75 m, then HI equals the elevation of the mark plus hi, or HI = 1521.75 m

Example 3.17. Suppose we set up a total station over a survey control marker and measure that the total station is 1.25 meters above the marker (i.e., hi is 1.25 m). We use the total station to measure a slope distance of 45.05 meters to some point of interest along with an accompanying angle of elevation θ of 13.92°. How high is the aim point above the marker?

Solution: Denote the slope distance as d_s, the height of the aim point above the total station as d_v, and the angle of elevation as θ. Then

$$\begin{aligned}\sin \theta &= d_v/d_s, \text{ so} \\ d_v &= d_s \cdot \sin \theta \\ &= 45.05 \cdot \sin(13.92°) \\ &= 10.84 \text{ m}\end{aligned}$$

Since the total station was 1.25 meters above the marker, the height of the aim point is 10.84 + 1.25 = 12.09 meters.
□

Example 3.18. Suppose we want to measure the height of a radio tower above the ground directly below the center of the tower (see Fig. 3.9 on the following page). Platforms between the tower's legs prevent measuring the height directly. We set up a total station beside the tower and measure the total station hi = 1.662 m. Then we set up a prism on a prism pole rod directly below the center of the tower and measure the hr, including prism, to be 2.025 m. The slope distance from the total station to the prism is measured at 47.141 m and the angle of elevation is 13°13'55" below horizontal. We then measure an angle of elevation of 43°39'30" and a slope distance of 63.430 m to the top of the tower using the reflectorless capability of the EDM. What is the distance from the top of the tower to the ground directly below?

Solution: The height of the tower (h) is the sum of three vertical distances: the hr, the distance from the prism to the height of the optical center of the total station's telescope ($h1$), and the

CHAPTER 3. REDUCTIONS AND COMPUTATIONS FOR PLANE SURVEYING

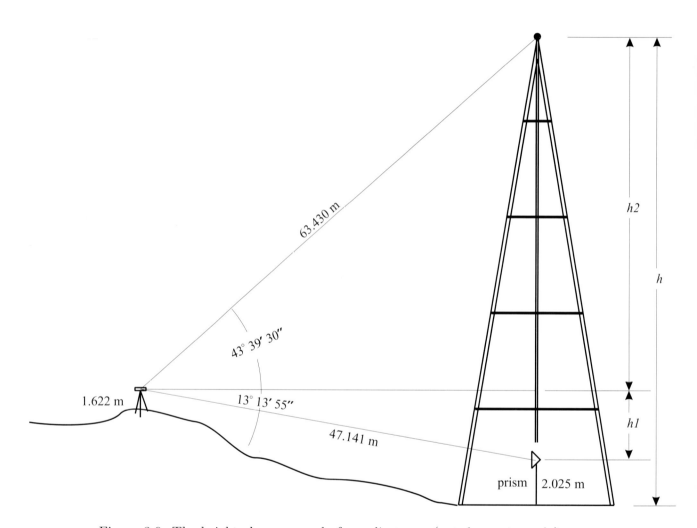

Figure 3.9: The height above ground of a radio tower (not drawn to scale).

3.6. REDUCTIONS FOR PLANE SURVEYING

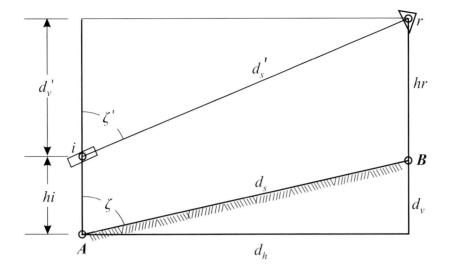

Figure 3.10: Reducing an observed zenith angle ζ' to the zenith angle ζ from A to B.

distance from the optical center of the total station's telescope to the top of the tower ($h2$). Using the definition of the sine

$$\sin(13°13'55'') = h1/47.141, \text{ so}$$
$$h1 = 47.141 \sin(13°13'55'')$$
$$= 10.790 \text{ m}$$

Similarly

$$\sin(43°39'30'') = h2/63.430, \text{ so}$$
$$h2 = 63.430 \sin(43°39'30'')$$
$$= 43.789 \text{ m}$$

So, $h = 2.025 + 10.790 + 43.789 = 56.604$ m. Notice that the height of the total station did not figure into the computation. Why not?
□

3.6.2 Reducing slope distances to vertical distances

Using the logic in section 3.6.1 on page 34, which showed how to reduce slope distances to horizontal distances, we can see that the formula relating slope distances to vertical distances is a simple modification of Eq. 3.9:

$$d_v = d_s \cos \zeta \tag{3.10}$$

where d_v is a vertical distance, d_s is a slope distance, and ζ is a zenith angle.

Figure 3.10 shows a total station located at the open circle at i, which is above station A. The height of the instrument is hi; and the height of the rod, including a prism reflector, is hr. The slope distance d'_s and zenith angle ζ' are observed, from which are derived horizontal distance

$d_h = d'_s \sin \zeta'$ and vertical distance $d'_v = d'_s \cos \zeta'$ from the instrument to the reflector at r. The height of B above A is then

$$d_v = hi + d'_v - hr$$
$$= hi + d'_s \cos \zeta' - hr \qquad (3.11)$$

Reducing slope distances to vertical distances is called **trigonometric leveling** (trig leveling), or **trigonometric heighting**. In practice, trig leveling is usually relegated to low accuracy work because (1) it ignores the Earth's curvature and (2) atmospheric refraction greatly limits observing zenith angles correctly. However, there are observation protocols that can be followed that allow trig leveling to be very accurate (Soycan 2006).

3.6.3 Reducing observed zenith angles down to the mark

The zenith angles (e.g., ζ) between the stations were not observed but they are needed to determine the actual slope distance between the markers. If $hi \neq hr$, then $\zeta' \neq \zeta$. Therefore, the observed zenith angle must be reduced down to the stations. It can be shown that the desired reduction is

$$\cot \zeta = \csc \zeta' \frac{hi - hr}{d'_s} + \cot \zeta' \qquad (3.12)$$

where cot is the cotangent (reciprocal of the tangent) and csc is the cosecant (reciprocal of the sine). Inspecting Eq. 3.12 reveals that if $hi = hr$, then the cosecant term goes to zero and the observed and reduced zenith angles are the same, as expected. Otherwise, the cosecant term acts to adjust the observed angle to account for the difference of the instrument and rod heights over the separation between the stations.

Example 3.19. If $\zeta' = 89°\ 14'\ 50''$, $hi = 1.653$ m, $hr = 2.105$ m, and $d'_s = 59.131$ m, find d_v and ζ.
Solution: Using $89°\ 14'\ 50'' = 89.247\,2222°$ and the other inputs in Eq. 3.11 gives

$$d_v = hi + d'_s \cos \zeta' - hr$$
$$= 1.653 + 59.131 \cos 89.247\,2222° - 2.105$$
$$= 0.325 \text{ m}$$

To find ζ

$$\cot \zeta = \csc \zeta' \frac{hi - hr}{d'_s} + \cot \zeta'$$
$$= \csc 89.247\,2222° \left(\frac{1.653 - 2.105}{59.131} \right) + \cot 89.247\,2222°$$
$$= 1/\sin 89.247\,2222° \left(\frac{1.653 - 2.105}{59.131} \right) + 1/\tan 89.247\,2222°$$
$$= 1.000\,09 \times (-0.007\,644\,04) + 0.013\,1392$$
$$= 0.005\,4945, \text{ so}$$
$$\tan \zeta = 1/0.005\,4945$$
$$\zeta = \arctan(1/0.005\,4945)$$
$$= 89°\ 41'\ 7''$$

□

3.6. REDUCTIONS FOR PLANE SURVEYING

Example 3.20. In this example the zenith angle exceeds 90°, meaning the line of sight to the target is downhill from the instrument. Given $\zeta' = 92°\,6'\,25''$, $hi = 1.5$ m, $hr = 2.105$ m, and $d'_s = 171.25$ m, find ζ.
Solution:

$$\cot\zeta = \csc\zeta' \frac{hi - hr}{d'_s} + \cot\zeta'$$

$$= \csc 92.106\,944\,44° \left(\frac{1.5 - 2.105}{171.25}\right) + \cot 92.106\,944\,44°$$

$$= 1/\sin 92.106\,944\,44° \left(\frac{1.5 - 2.105}{171.25}\right) + 1/\tan 92.106\,944\,44°$$

$$= 1.000\,676\,512 \times (-0.003\,532\,85) - 0.036\,789\,702\,32$$

$$= -0.040\,3249, \text{ so}$$

$$\tan\zeta = -1/0.040\,3249$$

$$\zeta = \arctan(-1/0.040\,3249)$$

$$= 92°\,18'\,33''$$

□

Surveying equipment companies sell adjustable prism poles that can be set so that the height of the instrument will equal the height of the rod. This eliminates the need for this reduction and the one in the next section. However, it is not always possible to set the two heights to be equal due to line-of-sight constraints.

3.6.4 Reducing observed slope distance to distance between the marks

In Fig. 3.10 on page 37, it is evident if $hi \neq hr$ then $d'_s \neq d_s$, meaning that the observed slope distance d'_s is not equal to the slope distance d_s between A and B unless the height of the instrument equals the height of the rod. It can be shown that the needed reduction is

$$d_s = \frac{d'_s \sin\zeta'}{\sin\zeta} \tag{3.13}$$

where ζ is the reduced zenith angle from Eq. 3.12.

Example 3.21. Given $\zeta' = 89°\,14'\,50''$, $\zeta = 89°\,41'\,7''$, and $d'_s = 59.131$ m, find d_s.
Solution: After converting to decimal degrees, from Eq. 3.13

$$d_s = \frac{d'_s \sin\zeta'}{\sin\zeta}$$

$$= \frac{59.131 \sin 89.247\,222\,222°}{\sin 89.685\,277\,78°}$$

$$= 59.127 \text{ m}$$

□

3.7 The Direct Problem in the Plane

3.7.1 Metes and bounds

The **direct problem** in the plane is the mathematical process of determining the planimetric coordinates of features of interest from bearings and horizontal distances. Bearings and distances appear on maps in the form of **metes and bounds** (Fig. 3.11 on the next page), defined by the Geodetic Glossary (NGS 2009) as "Those characteristics of a piece of land that are used in defining its boundary. The metes and bounds include measurements as well as monuments, etc. A 'metes and bounds description' of a piece of land is a complete description of its boundary, in which the distance (metes) and directions are given sequentially around the perimeter (bounds) of the piece of land." According to Merriam-Webster's Dictionary, a mete is akin to Old High German *mezzan* (to measure) and Latin *modus* (measure). "Bounds" refers to the physical demarcation of the boundaries of the property, such as iron pins driven into the ground, stakes driven into trees, or marks on monuments. The bearings and distances are observed during the traverse and used to determine positions by the direct problem, as discussed below.

3.7.2 Traversing

Coordinates of the origin

In a local survey, the coordinates of stations have no global meaning. The coordinates assigned to a local survey's origin are called **false eastings** and **false northings** because the station's absolute (geodetic) position is unknown. It is traditional to assign coordinates to the survey's origin that result in all coordinates in the survey being positive. Also, the value of the origin's easting is chosen to be different from the origin's northing by, say, an order of magnitude so that the false eastings and northings can be distinguished at a glance, such as $(1000, 10\,000)$.

Orienting a survey

Having chosen an origin, the surveyor must decide in which direction is north in order to construct a coordinate system. The chosen direction of north gives the survey its **orientation**. Suppose a total station was used to measure the place shown in Fig. 3.11 on the facing page. Point **A** was the POB. After the instrument was set up, the surveyor pointed at the azimuth mark, which is shown in the figure as the arrow pointing to **A**. The azimuth mark is displayed with the arrow symbol because that is the symbol stamped on brass cap surveying monuments set as azimuth marks. **Azimuth marks** are marks to which the azimuth from some primary station is either known or assumed to be known. Azimuth marks can be brass caps set in the ground but also can be very high objects, such as the red lights atop radio towers or the tops of skyscrapers, and the pinnacle of tall monuments (such as the Washington Monument) and church steeples. Such objects can be seen from far away and thus provide ideal targets for the initial backsight. However, in land surveying, one of the property bounds usually serves as an azimuth mark for the survey.

In the example in Fig. 3.11, upon pointing the instrument at the azimuth mark, the surveyor zeroed the instrument, meaning the horizontal circle was set to read $0°$. This established local "due north" to be the direction from **A** to the azimuth mark. In fact, the surveyor had no knowledge at all about the geodetic direction from **A** to the azimuth mark, so zero was chosen for lack of a better informed choice.

3.7. THE DIRECT PROBLEM IN THE PLANE

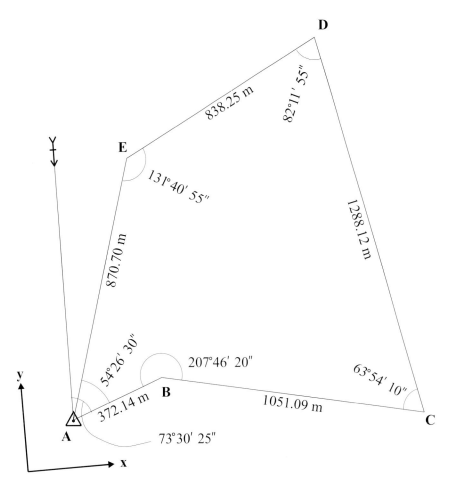

Figure 3.11: Traverse showing the measurements recorded by a total station. Notice that the local coordinate axes are perpendicular and parallel to the baseline from the monument to the azimuth mark, because the true azimuth from **A** to the azimuth mark is unknown in this case.

Traverse observations

Having zeroed on the azimuth mark, the surveyor pointed at **B** as the foresight. The first observations were the slope distance and bearing of (372.14 m, 73°30′25″). Moving to station **B**, station **A** became the backsight and station **C** became the foresight. From station **B** the observations were (1051.09 m, 207°46′20″). Notice that the angles are all measured clockwise from the backsight to the foresight, even if the observed bearing is greater than 180°. The traverse continues by occupying all the other stations in turn, repeating these observations until station **A** is observed as a foresight from **E**.

3.7.3 Positioning

After traverse observations have been collected, the next task is to compute coordinates from those observations with the goal of compiling a map. Using local Cartesian coordinates requires creating a local Cartesian coordinate system. This local horizontal coordinate system is called a **grid**, with **grid north** referring to the assumed direction for north.

Recall that a Cartesian coordinate is actually a linear offset from the origin (i.e., a straight-line distance along either the assumed north or east axis, with offsets in the north and east directions being positive and offsets in the south and west directions being negative). Offsets from the local origin along the east axis are called **eastings** (but not westings). Similarly, offsets from local origin in the assumed north/south direction are called **northings** (but not southings). The eastings and northings constitute **grid coordinates**.

The set of coordinates for a particular location is called a **position** and the process of computing positions from observations is called **positioning**. The basic idea behind positioning is the simple notion that the coordinates of the foresight equal the coordinates of the observation station plus a δe and δn equal to the horizontal offsets between the stations (see Fig. 3.12 on the next page). In particular, given the horizontal distance d_h and the azimuth α between the observing station and the foresight

$$\delta e = d_h \sin \alpha \tag{3.14}$$

and

$$\delta n = d_h \cos \alpha \tag{3.15}$$

These equations hold no matter what quadrant the foresight is in relative to the observing station because the sine and cosine functions produce correctly signed quantities that, when added to the coordinates of the observing station, always produce the correct offsets to the foresight. In Fig. 3.12, notice that δe is positive and that δn is negative, because **C** is to the east and south of **B**.

The following algorithm can be used to compute the stations' positions. Let $e_\mathbf{A}$ and $n_\mathbf{A}$ denote the easting and northing of station **A**, respectively, and $d_{\mathbf{A},\mathbf{B}}$ denote the horizontal distance from station **A** to station **B**. $\beta_{A,B,C}$ denotes the clockwise bearing at station **B** using station **A** as a backsight to station **C**.

Point of Beginning

The POB is handled as a special case because the bearing to the POB's backsight was either known or assumed, so it does not need to be calculated. In Fig. 3.11 on the preceding page, 73°30′25″ is the bearing from the POB to its backsight, and because that direction was assumed to be north, this bearing is also the azimuth from **A** to **B**. Given the observations in the figure and

3.7. THE DIRECT PROBLEM IN THE PLANE

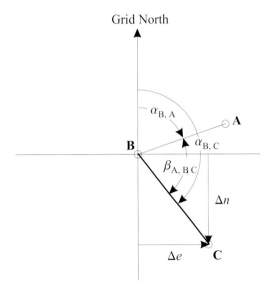

Figure 3.12: Angles and distances necessary to compute the coordinates of station **C**. In particular, notice that $\alpha_{B,C} = \alpha_{B,A} + \beta_{A,B,C}$, meaning the azimuth of the foresight from the observation station equals the azimuth of the backsight from the observation station plus the bearing from the backsight to the foresight. This is the key relationship among the angles observed in plane surveying.

assuming $\mathbf{A} = (1000, 10\,000)$, **B**'s easting is $e_\mathbf{A} + d_{\mathbf{A},\mathbf{B}} \sin \alpha_{\mathbf{A},\mathbf{B}} = 1000 + 372.14\ \sin(73°30'25'')$, and **B**'s northing is $n_\mathbf{A} + d_{\mathbf{A},\mathbf{B}} \cos \alpha_{\mathbf{A},\mathbf{B}} = 10\,000 + 372.14\ \cos(73°30'25'')$. Therefore, $\mathbf{B} = (1356.83, 10\,105.70)$.

Traverse points

Now let's consider the general case of stations in the survey without an azimuth mark. Suppose we are computing the coordinates from some station **U** to station **V** using station **T** as a backsight. We already have coordinates for **U** and **T** as a result of a previous traverse step.

1. Let $\alpha_{\mathbf{U},\mathbf{T}}$ denote the azimuth from **U** to **T**, the **back azimuth**. Compute $\alpha_{\mathbf{U},\mathbf{T}}$ from Eq. 3.8 as
$$\alpha_{\mathbf{U},\mathbf{T}} = \arctan[\delta e / \delta n] + \kappa \tag{3.16}$$
where $\delta e = \mathbf{V}_e - \mathbf{U}_e$ and $\delta n = \mathbf{V}_n - \mathbf{U}_n$. Pay careful attention to the order of the coordinates in these differences because reversing them will reverse the sign of the difference, which is incorrect. Get the correction for quadrant from Table 3.1 on page 33. $\alpha_{\mathbf{T},\mathbf{U}}$ was likely computed in the previous traverse step, so for plane surveying, $\alpha_{\mathbf{U},\mathbf{T}} = \alpha_{\mathbf{T},\mathbf{U}} + 180°$, which saves the trouble of computing the arctangent.

2. $\alpha_{\mathbf{U},\mathbf{V}} = \alpha_{\mathbf{U},\mathbf{T}} + \beta_{\mathbf{T},\mathbf{U},\mathbf{V}}$.

3. The coordinates of station **V** are computed as
$$e_\mathbf{V} = e_\mathbf{U} + d_{\mathbf{U},\mathbf{V}} \sin(\alpha_{\mathbf{U},\mathbf{V}}) \tag{3.17a}$$
$$n_\mathbf{V} = n_\mathbf{U} + d_{\mathbf{U},\mathbf{V}} \cos(\alpha_{\mathbf{U},\mathbf{V}}) \tag{3.17b}$$

Example 3.22. Referring to Fig. 3.11 on page 41, taking **B** as the observation station and **A** as the backsight, compute the coordinates of station **C**.

Solution: We were given the coordinates of the monument at station **A** as (1000, 10 000) and computed the coordinates at station **B** as (1356.83, 10 105.70). So, using Eq. 3.16, we compute $\alpha_{B,A}$ as

$$\begin{aligned}\alpha_{B,A} &= \arctan[\delta e/\delta n] + \kappa \\ &= \arctan[(e_A - e_B)/(n_A - n_B)] + \kappa \\ &= \arctan[(1000 - 1356.83)/(10\,000 - 10\,105.70)] + \kappa \\ &= \arctan[-356.83/-105.70] + \kappa \\ &= 73.4997° + \kappa\end{aligned}$$

From Table 3.1, κ is 180°, so $\alpha_{B,A} = 73.4997° + 180° = 253.5°$.

$\beta_{A,B,C} = 207°46'20'' = 207.772°$. $\alpha_{B,C} = \alpha_{B,A} + \beta_{A,B,C} = 253.5° + 207.772° = 461.272°$. Removing the angle in excess of 360°, $\alpha_{B,C} = 461.272° - 360° = 101.272°$.

Using Eq. 3.17a, the easting coordinate of station e_C is

$$\begin{aligned}e_C &= e_B + d_{B,C}\sin(\alpha_{B,C}) \\ &= 1356.82 + 1051.09\sin(101.272°) \\ &= 2387.63\end{aligned}$$

and the northing coordinate is

$$\begin{aligned}n_C &= n_B + d_{B,C}\cos(\alpha_{B,C}) \\ &= 10\,105.70 + 1051.09\cos(101.272°) \\ &= 9900.24\end{aligned}$$

Therefore, **C** = (2387.63, 9900.24). □

3.7.4 Checking computations and observations

When computing coordinates from a traverse it is fairly easy to make a mistake somewhere in the middle of what is often a long, complicated series of computations. Any mistake propagates throughout the remainder of the computational chain, ruining all subsequent work. However, checking the computations is easy if the traverse started and ended on a control point: simply compare the computed coordinates for the check point with its authoritative value. Traverses often form a closed loop. Checking against the coordinates of the starting point is very valuable, even if no egregious blunder occurred, because it lets you gauge the accuracy of your computations and measurements. Checking for errors and adjusting observations to enforce error constraints, such as loop closures, is a serious study in itself known as "observation adjustments" and is beyond the scope of this book (e.g., see Wolf and Ghilani 1997).

If the observations were without error in a closed-circuit (loop) traverse, the coordinates computed for the POB would exactly match the given (or assumed) coordinates. However, all observations have some amount of experimental error caused by imperfections in the instrument, by the environment, and by the surveyor. The planimetric separation between **A**'s computed position and its given position is the survey's **closure error**. **Closure precision** is obtained by normalizing the closure error by the traverse's length.

3.8. SAMPLING TRANSECTS

Example 3.23. The assumed coordinates of a survey's origin are $(e_O, n_O) = (1000, 10\,000)$ and its computed coordinates, denoted **POB**, are $(e_{POB}, n_{POB}) = (1000.005, 9999.975)$. Find the closure error and precision.
Solution: The closure error is

$$\begin{aligned}
\epsilon &= \sqrt{(e_O - e_{POB})^2 + (n_O - n_{POB})^2} \\
&= \sqrt{(1000.000\,\text{m} - 1000.005\,\text{m})^2 + (10\,000\,\text{m} - 9999.975\,\text{m})^2} \\
&= \sqrt{(-0.005\,\text{m})^2 + (0.025\,\text{m})^2} \\
&= \sqrt{0.000\,025\,\text{m}^2 + 0.000\,625\,\text{m}^2} \\
&= 0.026\,\text{m}
\end{aligned}$$

Suppose the length of the traverse was 845.605 m. Then the closure precision is $0.026/845.605 = 0.000\,03$. This is a very small number, so it is multiplied by one million to give a 30 parts per million closure precision. Alternatively, one can normalize the traverse length by the closure error as $845.60/0.0255 = 32\,523$ or about a 1:32 500 error, meaning there was one unit length of error for each 32 500 unit lengths of the traverse. Both methods of reporting mean the same thing.
□

3.8 Sampling Transects

Two examples of using a total station for natural resources fieldwork are presented. The first deals with laying out a transect for sampling vegetation. The second deals with creating stream profiles.

Natural resource managers often need to sample areas to determine statistics about their vegetation and wildlife populations. One of the tools used to do this is a sampling **transect**, which is a regular pattern of sampling points laid out in the study area. Mueller-Dombois and Ellenberg (1974, p. 93) describe a plot as "any 2–dimensional sampling area of any size. This includes quadrats, rectangular plots, circular plots, and belt transects (which are merely very long rectangular plots). Belt transects are often simply called strips or transects." For a population survey, a full inventory of the items of interest is conducted within each grid cell. The totals are used in statistical analyses to make estimates of populations and other facts of interest. For more details about sampling transects, see Cooperrider, Boyd, and Stuart (1986) and Rudnicki and Meyer (2007). Total stations, GPS receivers, and coordinate geometry are a perfect trio of tools for accurately laying out transects, although a tape and a compass often suffice.

3.8.1 Linear transects

A linear transects is simply a series of points spaced out evenly along a line whose endpoints have specified positions. Biologists use linear transects to sample biota that live close together, such as grasses. To conduct a population survey of grasses, a biologist might pick a study area and lay out some transects at random. Each transect would have the same length and the same number of sampling points. It is highly desirable to lay out the transects as accurately as possible to preserve the randomness of the sampling. One approach is to randomly generate the coordinates of the starting point and the orientation of the transect.

Suppose we are asked to lay out a transect, given the absolute coordinates of its starting point $(\mathbf{U_e}, \mathbf{U_n})$, its length d, its orientation α, and the distance between the sampling points d_s. Further suppose we have a GPS and a total station at our disposal.

Let us first assume we have a clear line of sight along the transect. We can use the GPS to find the starting point $\mathbf{U}$ by navigating to the specified coordinates, but neither the GPS nor the total station can establish the transect's orientation. We need to compute the coordinates of another point $\mathbf{V}$ along the line of sight of the transect using Eq. 3.14 and Eq. 3.15:

$$e_\mathbf{V} = e_\mathbf{U} + d \sin \alpha$$
$$n_\mathbf{V} = n_\mathbf{U} + d \cos \alpha$$

Any distance d will do; it need not be the length of the transect. Indeed, the farther $\mathbf{V}$ is from $\mathbf{U}$, the better. This is because a GPS has an expected accuracy better than three meters, and long baselines reduce the error in the orientation of the transect.

Example 3.24. With coordinates in meters, let $\mathbf{U} = (1523.65, 233\,519.23)$ and let $\alpha = 277.922\,54°$. Suppose we know that we can see 550 meters from $\mathbf{U}$. Find $\mathbf{V}$.
Solution:

$$\begin{aligned} e_\mathbf{V} &= 1523.65 + 550 \sin 277.922\,54° \\ &= 1523.65 + (-544.75) \\ &= 978.90 \end{aligned}$$

Similarly, $n_\mathbf{V} = 233\,595.04$ m. Therefore, $\mathbf{V} = (978.90, 233\,595.04)$.
□

Now, having the coordinates of both $\mathbf{U}$ and $\mathbf{V}$, we can use the GPS to place two stakes in the ground at these places. It may be possible to simply pull a string between them. If so, it may also be possible to use a tape measure to mark off the sampling points. However, obstructions such as shrubs might prevent measuring the intersampling distances directly. In this case we have two alternatives. First, we could compute the coordinates of the sampling points as just described, using the appropriate distance for d to place the points properly. However, this might be unacceptable because the GPS error is likely to be too large with respect to the sampling interval. If so, it's better to use the total station, which will be described next.

We have assumed that we have stakes in the ground at the starting point of a transect and at some other point on the line of the transect and that these points are intervisible. To stake out the sampling points with a total station, set up and level the instrument over the starting point. Next, we will point the instrument at the other point and zero the horizontal angle. Next, an assistant with a prism on a prism pole will be guided out to the approximate position of the first sampling point by the total station operator, to keep on the transect line. We will measure the distance to the prism with the total station and give the assistant instructions to guide him or her to the proper location of the first point. This process will be repeated until the point is set out properly.

The procedure given above assumes intervisibility of the sampling locations. This is often not the case, as with a transect laid out in a forest. In such a circumstance, we have to resort to a more complicated strategy. Suppose that we are to lay out a transect in which none of the stations are intervisibile but that we can set up our observation equipment off to the side of the transect and see all the stations. This might be the case if the transect is at the edge of a forest (Fig. 3.13).

3.9. YOUR TURN 47

Let us assume that we can see stakes in the ground at the starting point **U** of the transect and at a relatively distant point **V** elsewhere on the transect. We are told that the sample points are to be δs m apart, spaced evenly, starting at **U**. We will drive a third stake into the ground at a point **T** from where **U** and **V** are both visible. To stake out the sampling points,

1. Let $\beta_{\mathbf{U},\mathbf{T},\mathbf{V}}$ be the bearing measured at the total station from **U** to **V**. Let $d_{\mathbf{T},\mathbf{U}}$ and $d_{\mathbf{T},\mathbf{V}}$ be horizontal distances measured from the total station to **U** and **V**, respectively. Using the law of cosines, compute $d_{\mathbf{U},\mathbf{V}}$ and θ as shown in Fig. 3.13.

2. For each sample point s_i, let d_{s_i} be the distance from **U** to s_i, which is simply $i \times \delta s$. For example, if $\delta s = 10$ m and we're considering the third sample point s_3, then $d_{s_3} = 10 \times 3 = 30$ m.

3. With θ, d_{s_i} and $d_{\mathbf{T},\mathbf{U}}$, use the law of cosines to compute $d_{\mathbf{T},s_i}$, the horizontal distance from the total station to s_i.

4. With θ, $d_{\mathbf{T},s_i}$ and $d_{\mathbf{T},\mathbf{U}}$, use the law of cosines to compute β_i, the bearing measured at the total station from **U** to s_i.

5. Knowing $d_{\mathbf{T},s_i}$ and β_i, stake out s_i by zeroing the total station on **U** and directing the rod carrier to a distance $d_{\mathbf{T},s_i}$ in the direction β_i.

3.9 Your Turn

Problem 3.1. In Fig. 3.14 on the next page, what is α in radians?

Problem 3.2. In Fig. 3.14 on the following page, what is β in DMS?

Problem 3.3. The height of a tree is being measured with a total station. We measure the hi to be 1.36 m. We adjust the telescope so that it is level and measure the horizontal distance to the stem to be 29.19 m. We then tilt the telescope up to aim at the top of the crown and measure an AOE of 8.9664°. What is the height of the crown above the optical center of the total station's telescope?

Problem 3.4. Continuing problem 3.3, we notice that the base of the tree is downslope, so we tilt the telescope down and observe the angle from the horizontal to the ground at the base of the tree to be 0.5706°. What is the total height of the tree?

Problem 3.5. Suppose that in Fig. 3.5 on page 25, $\theta_{\mathbf{B}} = 148°\,35'\,40''$, $a = 509.25$ m, and $c = 308.09$ m. What is the length of b?

Problem 3.6. Suppose that in Fig. 3.5 on page 25, $a = 401.55$ m, $b = 655.12$ m, and $c = 521.41$ m. What are the three angles in DMS?

Problem 3.7. Suppose that in Fig. 3.5 on page 25, $\theta_{\mathbf{B}} = 138°\,05'\,20''$, $b = 95.41$ USFt, and $a = 55.09$ USFt. What is the length of c in meters?

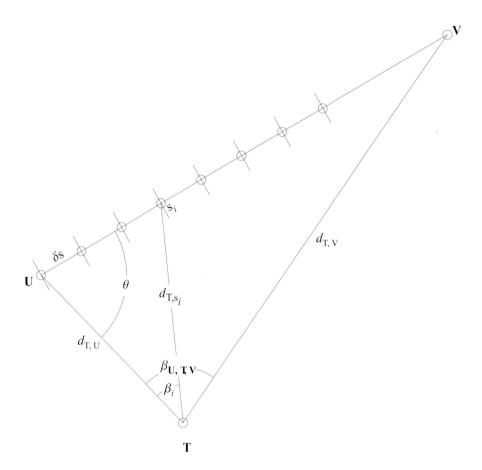

Figure 3.13: Staking out sampling locations for nonintervisible sites.

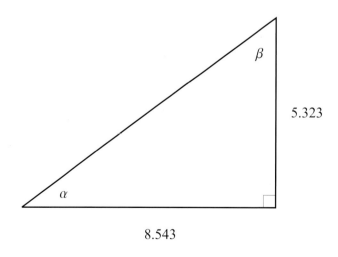

Figure 3.14: Right triangle.

3.9. YOUR TURN

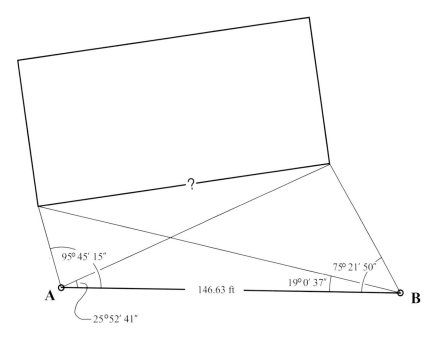

Figure 3.15: Determining the length of the façade of a building (not drawn to scale).

Problem 3.8. Suppose a total station has been set up at station **C** in Fig. 3.5, and $\theta_{\mathbf{C}}$ is observed to be $75°\,21'\,35''$. The observed slope distance to **B** is 142.91 USFt at a zenith angle of $90°\,1'\,40''$, and the slope distance to **A** is observed to be 100.52 USFt at a zenith angle of $88°\,14'\,00''$. What is the slope distance and zenith angle from **A** to **B**? What is the zenith angle from **B** to **A**?

Problem 3.9. Refer to Fig. 3.11 on page 41. What is the distance from point **B** to **E**, from point **B** to **D**, and from point **A** to **D**?

Problem 3.10. Suppose we want to determine the length of the front of a building with a façade too complicated to tape (see Fig. 3.15). We have at our disposal a theodolite but no EDM device. A previous survey established the separation between two monuments **A** and **B** in front of the building as being 146.63 USFt. With the theodolite we observe the angles shown in the figure. What is the length of the front of the building?

Problem 3.11. Using the dimensions indicated in Fig. 3.6 on page 27, calculate the area of the pasture.

Problem 3.12. Redraw Fig. 3.11 on page 41 by omitting the observed angles and including bearings for all the boundaries in the form N/S $<angle>$ E/W.

Problem 3.13. Compute the area of the parcel shown in Fig. 3.11.

Problem 3.14. We measure two features on a 1:10 000 map and find them to be separated by $d_m = 11.3$ cm. How far apart, in meters, are the features in the real world?

Problem 3.15. A map is published with an MSR of 1:24 000. Suppose two features are separated by $d_m = 4.95$ inches. How far apart are these features on the ground in feet?

50 CHAPTER 3. REDUCTIONS AND COMPUTATIONS FOR PLANE SURVEYING

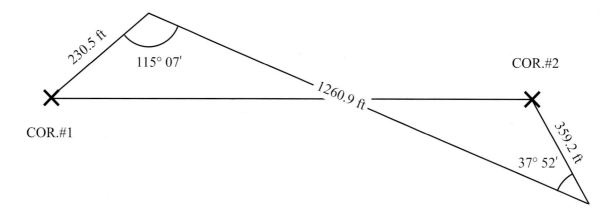

Figure 3.16: Corner occupation problem (based on Lindell, D. 2001. The Problem Corner, *Professional Surveyor* 21[10], 34).

Problem 3.16. Suppose a 1:24 000 map is measured with a scale whose least count is 0.01 in. What is the shortest ground distance in feet that can be resolved?

Problem 3.17. Suppose a 1:24 000 map is measured with a scale whose finest ruling is 1/32 of an inch. Two features are found to be 1 9/32 inches apart on the map. How far apart are they on the ground in feet?

Problem 3.18. Suppose an enlarged map's scale bar shows 1000 m, which is measured with a ruler to be 8.8 inches long. What is the enlarged map's scale ratio?

Problem 3.19. An aerial photograph has ground control panels whose centers have been surveyed determining that their centers are located at $\mathbf{A} = (391\,901.88, 808\,015.00)$ and
$\mathbf{B} = (391\,120.69, 808\,884.45)$, in meters. Find the planimetric distance between $\mathbf{A}$ and $\mathbf{B}$.

Problem 3.20. On a map with a coordinate grid, the closest horizontal grid line, with a northing coordinate of 4 250 000 m, is determined to be 49 meters on the ground south of some feature of interest. Another vertical grid line, with an easting coordinate of 425 000 m, is determined to be 881 meters west of the feature. If eastings increase to the east and northings increase to the north, what are the feature's coordinates?

Problem 3.21. Find the grid azimuth from $\mathbf{A}$ to $\mathbf{B}$ whose coordinates are as given in problem 3.19.

Problem 3.22. Find the grid azimuth from $\mathbf{B}$ to $\mathbf{A}$ whose coordinates are as given in problem 3.19.

Problem 3.23. Make a photocopy of Fig. 3.8 on page 32 without enlarging or shrinking the image. Determine the MSR of your photocopy. Determine the grid coordinates of points $\mathbf{B}$, $\mathbf{C}$, and $\mathbf{D}$.

Problem 3.24. In Fig. 3.16, surveyors could not occupy either corner to get a measured distance between them, so they measured the data shown. Determine the distance between the two corners (this problem is from Lindell 2001).

3.9. YOUR TURN

Problem 3.25. Compute the coordinates for all the stations in Fig. 3.11 on page 41. Assume the POB has coordinates of (1000 e, 10 000 n), and assume the azimuth from the POB to the azimuth mark is 0°. Be sure to check your work by computing coordinates for the POB at the end of the traverse (close the loop).

Problem 3.26. Field notes report that a total station was set up on station **A** with an hi of 1.45 m. A prism on a range pole was observed at station **B** at an hr of 2.15 m. A slope distance of 145.000 m was reported by the EDM. The zenith angle was observed to be 89°1′15″. Was the prism uphill or downhill of the telescope? What is the mark-to-mark zenith angle? Was **B** uphill or downhill from **A**?

Problem 3.27. Continuing problem 3.26, what is the mark-to-mark slope distance from **A** to **B**?

Problem 3.28. Field notes report that a total station was set up on station **M** with an hi of 1.26 m. A prism on a range pole was observed at station **N** at an hr of 2.15 m. A slope distance of 100.000 m was reported by the EDM. The zenith angle was observed to be 90°0′25″. Was the prism uphill or downhill of the telescope? What is the mark-to-mark zenith angle? Was **N** uphill or downhill from **M**?

Problem 3.29. Continuing problem 3.28, what is the mark-to-mark slope distance from **M** to **N**?

Problem 3.30. We desire to know the height of our total station (hi) for high-accuracy topographic work or for trigonometric leveling. To determine the hi, we set a stake with one corner higher than the others some distance away and, with differential leveling, determine that the change in height δH from the mark over which we've set up the total station to the high corner of the stake is +2.119 ft, meaning the high corner is 2.119 ft higher than the setup station mark. An assistant puts a stadia rod on the high corner and holds the rod so that it is vertical. With the total station, we observe a marking on the rod that reads 3.52 ft at a zenith angle $\zeta = 89°20′10″$, and at a slope distance d_s of 50.80 ft. What is the height of the instrument over the mark? What is the equation for hi given δH, ζ, and d_s?

Part II

Geometrical Geodesy

Chapter 4

Geographic Coordinates and Reference Ellipsoids

The existence of a science of geodesy, which is concerned with positioning across distances so great that the Earth's curvature cannot be ignored, raises the question under what circumstances plane surveying does not suffice. This question can be answered by an example that shows how the planar Earth model leads to incorrect results. Suppose a forward station is observed to be at a zenith angle of 90°, so it happens to be perfectly horizontal to the observation station. This implies that the height of the forward station equals that of the observation station because they are both in the same horizontal plane. However, the forward station is not at the same height as the observation station: in fact, it is higher (Fig. 4.1). The Earth's curvature causes its surface to drop away below the horizontal plane, so if the observation station is horizontal to the forward station, it must be above the observation station. This difference in height is the distance h in Fig. 4.1.

4.1 The Need for Geodetic Surveying

How quickly does h increase with distance c from the observation station? From Eq. 2.2, $c = a\gamma$, so $\gamma = c/a$. Then, from the definition of the cosine we have $\cos\gamma = a/(a+h)$, and after replacing γ with c/a, we solve for h:

$$\cos(c/a) = a/(a+h)$$
$$a + h = a\sec(c/a)$$
$$h = a\sec(c/a) - a \tag{4.1}$$

Given an assumption of a spherical Earth, this result is correct and exact, but it is probably not very intuitive because it gives the relationship between h and separation along the curved surface c in terms of the secant function. Mathematical relationships involving trigonometric functions arise frequently in geodesy, and we often want to reexpress those relationships without the functions. A common way to do this is to replace the function by its polynomial representation. Trigonometric functions require polynomials with an infinite number of terms, so such polynomials are called **infinite series** (see any standard calculus textbook for a treatment of infinite series). The infinite series for the secant of x is

$$1 + x^2/2 + 5x^4/24 + 61x^6/720 + 277x^8/8064 + 50\,521x^{10}/3\,628\,800 + \ldots$$

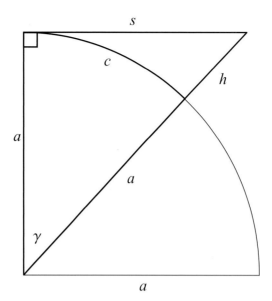

Figure 4.1: Trading height for position: assuming a flat Earth makes distant objects appear lower and farther away than what is real.

If $0 \leq |x| < 1$, then the contribution of the terms whose exponents are greater than one (called "higher-order terms") becomes very small very quickly. This allows us to safely simplify the polynomial by omitting certain higher-order terms. In this case, we'll omit terms higher than second order and see how large an error this introduces. Substituting c/a for x and replacing the secant in Eq. 4.1 with the truncated infinite series gives

$$\begin{aligned} h &= a \sec(c/a) - a \\ &\approx a\left(1 + (c/a)^2/2\right) - a \\ &\approx a + a(c/a)^2/2 - a \\ &\approx a\,(c/a)^2/2 \\ &\approx c^2/2a \end{aligned} \tag{4.2}$$

where c represents the distance between two places and a represents the radius of the curved surface representing the Earth. We can rearrange Eq. 4.2 to determine the station separation c for $h = 1$ cm: solving for c, and setting $a = 6\,371\,000$ m[1] and $h = 0.01$ m gives $c \approx 357$ m. This is a rather egregious error for such a short station separation. A similar analysis for the horizontal distance exaggeration $s - c$ shows that

$$s - c = a\,\tan(c/a)$$

Substituting its infinite series for the tangent function leads to $s - c \approx c^3/3a^2$. This function grows more slowly than Eq. 4.2 for short distances because $a^2 \gg a$ (Fig. 4.2). The horizontal error reaches 8 mm over 10 km, whereas the vertical error is 8 m over the same distance. This illustrates why plane surveying suffices for mapping relatively small areas but becomes inadequate for relatively long distances.

---

[1] See Jekeli (2005, p. 2–23) for a discussion about the mean Earth radius.

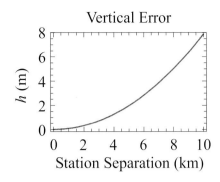 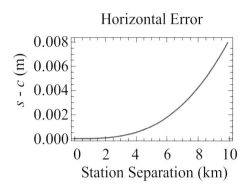

Figure 4.2: The left panel shows vertical error caused by a plane approximation. The right panel shows horizontal error.

4.2 Ellipses and Ellipsoids of Revolution

Abandoning plane surveying requires adopting some surface other than a plane upon which to perform computations such as inversing and the direct problem. This alternate surface should be mathematically simple enough so that geodetic computations are not overly complicated. Moreover, it should closely resemble the Earth's shape, which is nearly spherical but not perfectly so. One important part of geometrical geodesy is to create mathematical surfaces that accurately reflect the shape and size of the Earth – and its gravity field, it turns out.

A circle happens to fit the Earth's profile in the equatorial plane very well. However, the Earth's diurnal (daily) rotation causes its mass to bulge out near the equator compared to the poles. As a consequence, a meridian is fit better by an oblate ellipse (longer in the equatorial plane; like the profile of a American football) than by a circle. Rotating an ellipse about its longer axis results in an ellipse of revolution; if the ellipse is oblate, then the figure is an **oblate ellipsoid**, and if the ellipse is prolate (longer perpendicular to the equatorial plane; egg shaped), then the figure is a **prolate ellipsoid**. A meridian-shaped oblate ellipsoid is a more realistic Earth model than a sphere, so an oblate ellipsoid is the figure geodesists generally prefer to be the representation the Earth's macroscopic shape. An ellipsoid constructed for the purposes of geometric geodesy is called a **reference ellipsoid**.

Ellipses have a **major** and a **minor** axis, being the longer and shorter axes of the ellipse, respectively. The distances from an ellipse's center to the ellipse participate frequently in geometrical geodesy computations, so they are given their own names and symbols. The length of the **semimajor** axis, denoted a, is the distance from an ellipse's center to the ellipse along the major axis (see Fig. 4.3 on the next page); the length of the major axis is $2a$. Similiarly, the length of the **semiminor** axis, denoted b, is the distance from an ellipse's center to the ellipse along the minor axis; the length of the minor axis is $2b$. For example, the semimajor axis of the GRS 80 reference ellipsoid is $a = 6\,378\,137$ m, exactly, and the semiminor axes of the GRS 80 reference ellipsoid is $b = 6\,356\,752.314$ m (Moritz 2000).

The equation of an ellipse takes a very simple form when expressed using the semimajor (X) and semiminor (Z) axis coordinates:

$$\frac{X^2}{a^2} + \frac{Z^2}{b^2} = 1$$

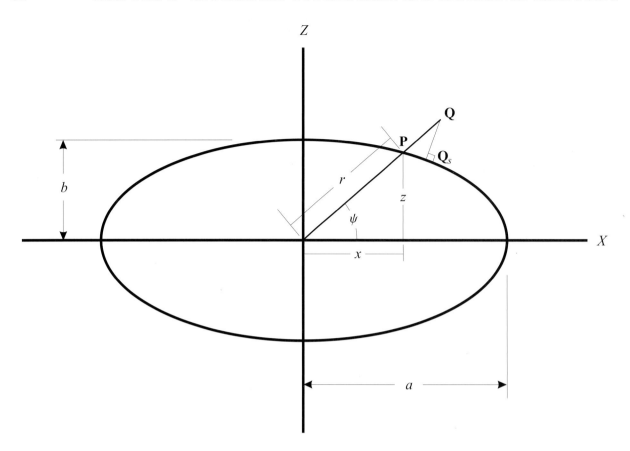

Figure 4.3: Ellipse parameters. The length of the semimajor axis is a, and the semiminor axis is b. Geocentric latitude is ψ. The distance to the center of the ellipse is r.

We use Z instead of Y because the minor axis will be the Z-axis in the Earth-Centered, Earth-Fixed coordinate system, to be discussed in section 5.1.

4.3 Latitude and Longitude

It was natural to use linear offsets on a planar reference surface for plane surveying coordinates, and it is likewise natural to use angular offsets for a curved surface. The angles chosen for this task are longitude and latitude.

Latitude and longitude are known as **geographic coordinates**, meaning they establish the location of some place on the Earth. Geographic coordinates can be natural or geometric (Kovalevsky et al. 1989, p. 145). **Natural coordinates** depend upon observations of natural phenomena, such as astronomic observations of star positions. Astronomically determined coordinates are known as **astronomic coordinates** (Moffitt and Bossler 1998, ch. 12). **Geometric coordinates** refer to some mathematical reference system and do not depend on observations of physical quantities. No knowledge of, say, the Earth's gravity field is needed to work with geometric coordinates. This independence from physical phenomena imparts a certain immutability to geometric coordinates, at least in principle, which is highly desirable. Natural coordinates are transformed into geometric

4.3. LATITUDE AND LONGITUDE

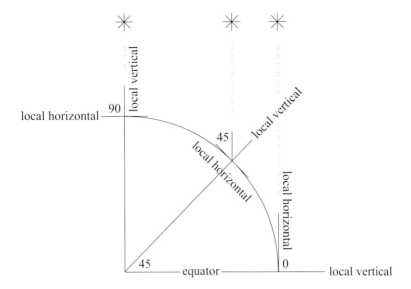

Figure 4.4: Astronomic latitudes at various places. The image shows one quadrant of the Northern Hemisphere as seen looking from a place in space in the equatorial plane. Polaris is so far away that the light coming from it effectively arrives everywhere on Earth as parallel rays.

coordinates by applying reductions, which will be discussed throughout the remainder of this book.

4.3.1 Astronomic latitude and longitude

Astronomic latitude is the angle between the local horizon and the zenith direction as determined by observing celestial bodies (see Fig. 4.4). Historically, measuring latitude to an accuracy adequate for maritime navigation was fairly simple in the Northern Hemisphere due to the fortunate existence of a North Star, which is currently Polaris. During the last 3000 years, the Earth's rotational axis has been oriented such that Polaris appears to be at a fixed position in the northern night sky, at least to the eye.[2] As shown in Fig. 4.4, if one were to view Polaris from the North Pole, Polaris would appear directly overhead, at an angle of 90° above the horizon, at the zenith. If one were to view Polaris from a place on the Earth halfway between the North Pole and the equator, Polaris would appear at an angle of 45° above the horizon. And, if viewed from the equator, Polaris would appear on the horizon. This relationship is the basis of latitude, which varies from 90° at the poles to 0° at the equator, thus making the equator the natural datum for latitude. By convention, positions in the Northern Hemisphere have positive latitudes and positions in the Southern Hemisphere have negative latitudes. Thus, the North Pole has a latitude of 90°, the equator has a latitude of 0°, and the South Pole has a latitude of −90°. It is customary to denote northern latitudes with the letter N (e.g., 47.5° N) and southern latitudes with the letter S (e.g., 21.35° S). The symbol for astronomic latitude is Φ (uppercase phi).

Measuring longitude was historically far more difficult than latitude (Sobel and Andrewes 1995). Polaris provided a natural fixed point from which to measure latitude, but a rigid body can rotate about only one axis at a time, so it is impossible for there to also be an "East Star" to serve as a

---

[2]Earth's rotational axis undergoes a slow circular movement (see section 7.1.2).

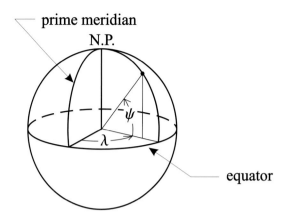

Figure 4.5: Geocentric latitude and longitude. Geocentric latitude is denoted by ψ. Geocentric longitude is denoted by λ.

measuring point for longitude. Ultimately, the problem of measuring longitude would be solved by measuring time: the Earth rotates through 360° of longitude in 24 hours, so a one-hour separation in time implies a $360°/24 = 15°$ separation in longitude. Clocks have existed since antiquity, but a spring-driven chronometer (as opposed to one that is pendulum driven, whose swinging could be affected by, say, the motion of a ship on the waves) would not be available until the 18th century. Although longitude can be determined by celestial events, it has usually been determined by time differences up to the advent of global navigation satellite systems. Λ (uppercase lambda) is the symbol for astronomic longitude.

4.3.2 Geocentric latitude and longitude

Geocentric latitude is the angle between a line segment connecting a reference ellipsoid's center to a location of interest, and the projection of that segment into the equatorial plane (Fig. 4.5). The symbol ψ (psi) stands for geocentric latitude.

A **meridian** is a curve formed by the intersection of a reference ellipsoid and a plane containing the ellipsoid's semiminor axis. The prime meridian is the meridian from which the other meridians are reckoned and assigned a longitude coordinate of 0°. The prime meridian is usually[3] taken to be the meridian passing through the Royal Observatory, Greenwich, England. **Geocentric longitude** λ is an angle in the equatorial plane from the prime meridian to some meridian of interest, increasing positive to the east.

Longitudes require special attention because western longitudes are denoted two different ways (Fig. 4.7). Meridians in the Western Hemisphere are denoted as angles east or west of the prime meridian. For example, the meridian in the Western Hemisphere halfway between the international date line (180° E) and the prime meridian is denoted equally correctly as either 270° E or 90° W. Longitudes expressed as western angles should be thought of as being mathematically negative because they are being reckoned in the direction opposite the definition: recall that longitude is reckoned positive to the east. To write a longitude as a western longitude, simply subtract 360°

---

[3]This is purely a convention. The prime meridian has been located in different places by different geodesy groups. For example, it has also been placed in Washington, D.C., and in Paris.

4.4. REFERENCE ELLIPSOIDS

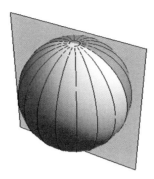

Figure 4.6: Meridians are lines of constant longitude formed by the intersection of the ellipsoid and a plane containing the ellipsoid's minor axis. For a reference ellipsoid, meridians are ellipses.

from its value as an eastern longitude. This will result in a negative value, indicating an angle in the Western Hemisphere.

Example 4.1. What is 331° 35′ 29″ E written as a western longitude?
Solution: First, convert to DD: $331°35'29''$ E = $331.591\,3889°$ E. Second, subtract 360°: $331.591\,3889° - 360° = -28.408\,6111°\,\text{W} = 28°24'31''\,\text{W}$.
□

A line of constant latitude forms a circle on a reference ellipsoid's surface that is parallel to the equatorial plane and whose center coincides with the minor axis. Any two such circles are parallel because the planes that contain them are parallel. Consequently, lines of latitude are called **parallels** (Fig. 4.8 on the next page).

Neither latitudes nor longitudes are commonly written as signed quantities and with a letter indicating a hemisphere, because there is some possible confusion in doing so. Consider a longitude of $-48°$ W. Strictly speaking, this is the same as 48° E because the minus sign and the W essentially create a double negative. However, it's extremely unlikely that anyone would denote 48° E in such an awkward way. Upon seeing $-48°$ W, it seems that a mistake may have been made, that the angle perhaps should have been 48° W. However, we cannot be sure. Therefore, geographic coordinates are either written with an arithmetic sign or with a letter indicating a hemisphere, but not both.

Figure 4.9 shows the system of parallels and meridians. Note that the parallels intersect the meridians at right angles, just like a grid.

4.4 Reference Ellipsoids

Until recently, the shapes and sizes of reference ellipsoids were established from extensive triangulation networks – a collection of numerous monumented, intervisible stations (triangulation stations) from which the angles between the stations could be observed (Gore 1889; Crandall 1914; Shalowitz 1938; Schwarz 1989; Dracup 1995; Keay 2000). There were at least two different methods used

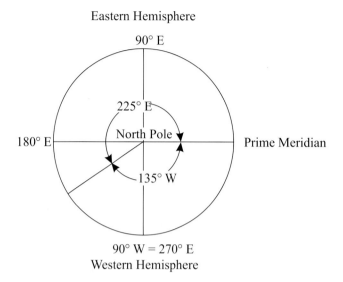

Figure 4.7: Looking at the North Pole from above, a particular meridian in the Western Hemisphere is equally well denoted as 135° W or 225° E.

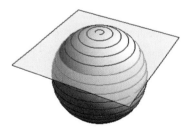

Figure 4.8: Parallels are lines (circles) of constant latitude formed by the intersection of the ellipsoid and a plane perpendicular to the ellipsoid's minor axis and, thus, parallel to its equator.

4.4. REFERENCE ELLIPSOIDS

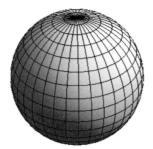

Figure 4.9: Grid of parallels and meridians. Parallels and meridians together form a coordinate system similar to a Cartesian grid of lines parallel to the x- and y-axes.

to create an ellipsoid: (i) the arc method, which was used for the Airy 1830, Everest 1830, Bessel 1841 and Clarke 1866 ellipsoids; and (ii) the area method that was used for the Hayford 1909 ellipsoid. The lengths of (at least) one starting baseline and one ending baseline were measured with instruments such as rods, chains, wires, or tapes, and the lengths of the edges of the triangles were subsequently propagated through the network mathematically by triangulation. For early triangulation networks, vertical distances were used for reductions and typically came from trigonometric heighting or barometric measurements, although for NAD 27, "a line of precise levels following the route of the triangulation was begun in 1878 at the Chesapeake Bay and reached San Francisco in 1907" (Dracup 1995). The ellipsoids deduced from triangulation networks were, therefore, custom-fit to the locale in which the survey took place. The result of this was that each region in the world thus measured had its own ellipsoid, and this gave rise to a large number of them; see NIMA (1997) and Table 4.1 for the parameters of many ellipsoids. It was impossible to create a single, globally applicable reference ellipsoid with triangulation networks due to the impossibility of observing stations separated by large bodies of water.

There is longitude associated with references ellipsoids: **geodetic longitude** is the angle in the equatorial plane of a reference ellipsoid between the prime meridian and the meridional plane of the point of interest reckoned positive to the east. The ISO/IEC18026:2006(E) and ISO19111 both accepted λ (lambda) as the symbol for longitude, both geodetic and geocentric. Geodetic longitude equals geocentric longitude because oblate ellipsoids have circular (not elliptical) equators.

There are several latitudes associated with reference ellipsoids, as will be discussed in section 4.5 on page 71.

4.4.1 Reference ellipsoid parameters

Geodetic positioning requires computing angles and distances on the surface of a reference ellipsoid. These tasks require a number of ellipsoidal parameters (see Ewing and Mitchell 1970 and Rapp 1989a for derivations of the following formulæ).

Table 4.1: Geodetic reference ellipsoid parameters. $1/f$ is the reciprocal flattening.

Semimajor (m)	$1/f$ or semiminor (m)	Ellipsoid
6377563.396	6356256.909	Airy 1830
6377104.43	300.0	Andrae 1876 (Denmark included)
6378137.0	298.25	Applied Physics 1965
6378160.0	298.25	Australian National Spheroid
6377397.155	299.1528128	Bessel 1841
6377483.865	299.1528128	Bessel 1841 (Namibia)
6378206.4	6356583.8	Clarke 1866
6378249.145	293.465	Clarke 1880 modified
6375738.7	334.29	Commission des Poids et Mesures 1799
6376428.	311.5	Delambre 1810 (Belgium)
6378136.05	298.2566	Engelis 1985
6377298.556	300.8017	Everest (Sabah and Sarawak)
6377276.345	300.8017	Everest 1830
6377304.063	300.8017	Everest 1948
6377301.243	300.8017	Everest 1956
6377295.664	300.8017	Everest 1969
6378166.	298.3	Fischer (Mercury Datum) 1960
6378150.	298.3	Fischer 1968
6378137	298.257222101	GRS 80 (IUGG 1980)
6378160.0	247.2471674273	GRS 67 (IUGG 1967)
6378388	297	Hayford 1909
6378200	298.3	Helmert 1906
6378270	297	Hough 1960
6378140.0	298.257	IAU 1976
6378160	298.247	Indonesian 1974
6378388	297	International 1924
6378163.	298.24	Kaula 1961
6378245	298.3	Krassovsky 1940
6378139.	298.257	Lerch 1979
6397300.	191.0	Maupertius 1738
6377340.189	6356034.448	Modified Airy
6378155	298.3	Modified Fischer 1960
6378137.0	298.257	MERIT 1983
6378145.0	298.25	Naval Weapons Laboratory 1965
6378157.5	6356772.2	New International 1967
6376523.	6355863	Plessis 1817 (France)
6378160	298.25	South America 1969 Spheroid
6378155.0	6356773.3205	Southeast Asia
6378136.0	298.257	Soviet Geodetic System 1985 (SGS 85)
6376896.0	6355834.8467	Walbeck
6378165	298.3	WGS 60
6378145	298.25	WGS 66
6378135	298.26	WGS 72
6378137	298.257223563	WGS 84

4.4. REFERENCE ELLIPSOIDS

Flattening and eccentricity

The size of an ellipse is set by the length of its semimajor axis, whereas its shape is set by some measure of its noncircularity. Noncircularity can be expressed three ways: semiminor axis, flattening, or eccentricity. Denoting the lengths of the semimajor and semiminor axes as a and b, as usual, **flattening** is defined as

$$f = (a-b)/a = 1 - b/a \tag{4.3}$$

From Eq. 4.3, $0 \leq f \leq 1$. If $a = b$, then $f = 0$; i.e., there is no flattening if the semimajor and semiminor axes have the same length, as intuition would suggest. The Earth is nearly spherical, so a and b have nearly the same value (Table 4.1). Consequently, f is a small number, around $1/3$ of 1%, so it is more convenient to report the reciprocal of the flattening: $1/f$. For example, the flattening of the GRS 80 ellipsoid is $f = 0.003\,352\,810\,681\,18$, which gives $1/f = 298.257\,222\,101$.

Eccentricity can be computed in terms of flattening as

$$\epsilon = \sqrt{2f - f^2}$$

ϵ (epsilon) is the symbol for eccentricity both in common literature and as adopted by the International Organization for Standardization (ISO) and the International Electrotechnical Commission (IEC) in ISO/IEC18026:2006(E), although other authors have used different symbols, such as e. In this book, e denotes a coordinate in the East-North-Up coordinate system, so in addition to being a standard, using ϵ for eccentricity avoids imbuing e with multiple meanings. There is another form of eccentricity, so ϵ is sometimes called the "first eccentricity."

The square of the eccentricity appears in formulæ more often than eccentricity itself, so it is usually given in its squared form:

$$\epsilon^2 = 2f - f^2 = \frac{a^2 - b^2}{a^2} = (a^2 - b^2)/a^2 \tag{4.4}$$

From Eq. 4.4, $0 \leq \epsilon^2 \leq 1$.

Example 4.2. Compute the the square of the first eccentricity of the GRS 80 ellipsoid.
Solution: Geodetic computations are often used to compute coordinates with twelve significant digits. Millimeter-accuracy computations require using the parameters in Table 4.1 exactly as given: for example, do not omit digits from the GRS 80 flattening value when computing eccentricity, and then carry all those digits throughout a computation. Use the memory in your hand calculator to store intermediate values, if necessary. The square of the first eccentricity of the GRS 80 ellipsoid is

$$\begin{aligned}\epsilon^2 &= 2f - f^2 \\ &= 2 \cdot 0.003\,352\,810\,681\,18 - 0.003\,352\,810\,681\,18^2 \\ &= 0.006\,694\,380\,022\,90\end{aligned}$$

□

The second eccentricity $(\epsilon')^2$ of an ellipse is defined in terms of the first eccentricity as

$$(\epsilon')^2 = \epsilon^2/(1 - \epsilon^2) = (a^2 - b^2)/b^2 \tag{4.5}$$

Example 4.3. Compute the square of the second eccentricity of the GRS 80 ellipsoid.
Solution:

$$\begin{aligned}(\epsilon')^2 &= \epsilon^2/(1-\epsilon^2) \\ &= 0.006\,694\,380\,022\,90/(1-0.006\,694\,380\,022\,90) \\ &= 0.006\,739\,496\,775\,48\end{aligned}$$

□

4.4.2 Distance from an ellipse to its center

The distance r from an ellipse to its center changes as a function of the geocentric latitude ψ, being equal to the semimajor axis at 0° and the semiminor axis at 90°. The distance r is given by Eq. 3.53 in Rapp (1989a, p. 23) as

$$r = b/\sqrt{1-\epsilon^2 \cos^2 \psi} \qquad (4.6)$$

Notice that, with the exception of $\psi = 0°$ and $\psi = 90°$, r is not perpendicular to the ellipse (Fig. 4.3 on page 58). Consider point **Q** which is "above" **P** in the sense that **Q** lies on the line segment from the center to **P** but is farther from the center than **P**. However, **Q** is not "vertically above" **P**, because the line segment is not perpendicular to the ellipse. Instead, **Q** is "vertically above" **Q**$_\mathbf{s}$, which has a different geocentric latitude than **P**. It is for this reason that geocentric latitude is not the type of latitude used to describe positions on the Earth, because if we did, changing a point's height would also change its latitude, which would ordinarily be considered nonsensical. An acceptable alternative can be found from radii of curvature.

4.4.3 Radii of curvature

Radius of curvature is a local measure of how "tight" a portion of a curve is (see Casey 1996 or McCleary 1994 for detailed discussions of curvature). The radius of curvature R of a curve of the form $y = f(x)$ possessing at least second-order derivatives is

$$R = \frac{[1+(dy/dx)^2]^{3/2}}{d^2y/dx^2}$$

A circle that is tangent to a curve at a point on the curve such that the circle's radius exactly matches that of the curve at that point is called an **osculating circle**. The reciprocal of the radius of an osculating circle is the **radius of curvature** of the curve at that point. Since a straight line has no curvature, the osculating circle fit to a straight line must have an infinitely large radius, and its reciprocal can be said to be zero. A tight bend in a curve would have large curvature. A tight bend would be fit by a small osculating circle, and the smaller the osculating circle, the larger the radius of curvature.

Radii of curvature play an important role in determining geodesic distances. Distance formulæ on curved surfaces require a parameter indicating the radius of the surface upon which the computation is to be performed. With a spherical model of the Earth, that number is simply the radius of the sphere. An ellipsoid, however, has three types of radii of curvature: those in a meridional plane, those in a plane oriented east-west, and those in a plane with arbitrary orientation.

4.4. REFERENCE ELLIPSOIDS

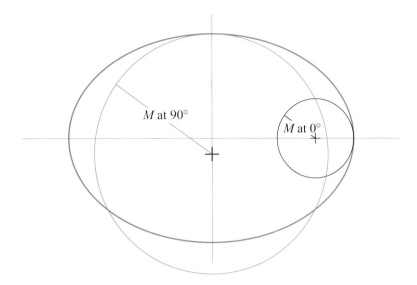

Figure 4.10: Radius of curvature in the meridian for $\phi = 0°$ and $\phi = 90°$.

Radius of curvature in the meridian

The **radius of curvature in the meridian** M is the radius of curvature of an ellipsoid in the plane of a meridian. It is the radius to use in geodesic length computations in the special case that the geodesic falls exactly along a meridian (see section 6.1.8 on page 113). M varies with latitude, achieving a maximum at the equator and a minimum at the poles for an oblate ellipse. The equation for the radius of curvature in the meridian is

$$M = \frac{a(1 - \epsilon^2)}{(1 - \epsilon^2 \sin^2 \phi)^{3/2}} \tag{4.7}$$

where a is the length of the semimajor axis, ϵ^2 is from Eq. 4.4, and ϕ is geodetic latitude (section 4.5 on page 71). Note that M at $0° = b^2/a < b$ and M at $90° = a^2/b > b$; the inequalities hold when $\epsilon > 0$ (Fig. 4.10).

Example 4.4. Compute the GRS 80 radius of curvature in the meridian at a geodetic latitude $\phi = 41.980\,97°$.
Solution:

$$\begin{aligned} M &= \frac{a(1 - \epsilon^2)}{(1 - \epsilon^2 \sin^2 \phi)^{3/2}} \\ &= \frac{6\,378\,137\,(1 - 0.006\,694\,380\,022\,90)}{(1 - 0.006\,694\,380\,022\,90\,\sin^2 41.980\,97°)^{3/2}} \\ &= 6\,364\,009. \end{aligned}$$

The solution has been rounded to the nearest meter.
□

Example 4.5. What is the length on the GRS 80 reference ellipsoid of 1 second of arc along a meridian at the equator?

Solution: The curvature of reference ellipsoids is nearly constant over such a small distance as 1 arc second, so they can be approximated as having a constant radius. The arc is along a meridian that can be approximated as a circle of constant radius equal to the radius of curvature in the meridian. From Eq. 2.2, $d = r\theta$. Letting r be M at $0° = 6\,335\,439.327$ m and $\theta = 1''$, after converting θ to radians, we get $d = 30.715$ m.
□

Example 4.6. Problem: What is the length on the GRS 80 reference ellipsoid of 1 second of arc along a meridian at the North Pole?
Solution: Letting r be M at $90° = 6\,399\,593.626$ m and $\theta = 1''$, we get $d = 31.026$ m.
□

Radius of curvature in the prime vertical

As shown in Fig. 4.11 on the facing page, the **radius of curvature in the prime vertical** N is the length of a line segment normal (perpendicular) to an ellipsoid, extending between the ellipsoid's surface and its minor axis (see Bowring 1987 for a simple proof). This line segment is parallel to the vector **v** normal to the ellipsoid's surface at the point of interest. The formula for the radius of curvature in the prime vertical is

$$N = \frac{a}{(1 - \epsilon^2 \sin^2 \phi)^{1/2}} \qquad (4.8)$$

where a is the length of the semimajor axis, ϵ^2 is eccentricity from Eq. 4.4, and ϕ is geodetic latitude. Note that N at $0° = a$ and $N = M$ at $90° = a^2/b > b$. Also, **v** is not parallel to the line segment connecting the ellipsoid's surface with its center (shown in green in Fig. 4.11).

Example 4.7. For the GRS 80 ellipsoid, at a geodetic latitude $\phi = 41.980\,97°$, compute the radius of curvature in the prime vertical.
Solution:

$$\begin{aligned} N &= \frac{a}{(1 - \epsilon^2 \sin^2 \phi)^{1/2}} \\ &= 6\,378\,137/(1 - 0.006\,694\,380\,022\,90 \sin^2 41.980\,97°)^{1/2} \\ &= 6\,387\,710. \end{aligned}$$

The solution has been rounded to the nearest meter.
□

Example 4.8. What is the length on the GRS 80 reference ellipsoid of one second of arc along the equator?
Solution: The equator is a circle of constant radius equal to the radius of curvature in the prime vertical at $\phi = 0°$, which equals the semimajor axis. So, from Eq. 2.2, $d = r\theta$. Letting $r = N$ at $0°$ gives $r = 6\,378\,137$ m. Setting $\theta = 1''$, and after converting θ to radians, we get $d = 30.922$ m.
□

4.4. REFERENCE ELLIPSOIDS

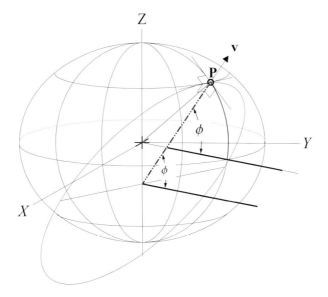

Figure 4.11: Geodetic latitude is ϕ. The vector normal to the surface of the ellipsoid is **v**. The inverse of the length of the dashed red line is the radius of curvature in the prime vertical N at **P**.

Radius of curvature in the normal section

A **section** is a curve formed by the nonempty intersection of a figure and a plane. A section of an ellipsoid must either be a point (the plane is tangent to the ellipsoid), a circle (a parallel of latitude), or an ellipse (the general case). A **normal section** is a curve formed by the intersection of a figure and a plane containing at least one surface normal vector of the ellipsoid (see Fig. 6.3 on page 110 and section 6.1.6). Normal sections are specified in geodesy by a point on the normal section and an azimuth specifying the direction of the sectioning plane. A normal section is the curve formed by the intersection of a reference ellipsoid and a plane (the sectioning plane) that contains the ellipsoid's surface normal vector at the specified point and that is oriented according to the specified azimuth.

The osculating circle for M is contained in a meridional plane, and thus is a normal section aligned north-south; think of the sections of an orange. The osculating circle for N is perpendicular to that of M, and thus is a normal section aligned east-west at their common point (**P** in Fig. 4.11). In general $M \neq N$, so it is evident that the radius of curvature on an ellipsoid varies with azimuth. The radius of curvature in the normal section R_α at some point on an ellipsoid is a combination of M and N. Its formula, which is known as Euler's theorem (Ewing and Mitchell 1970; Casey 1996), is

$$\frac{1}{R_\alpha} = \frac{\cos^2 \alpha}{M} + \frac{\sin^2 \alpha}{N} \qquad (4.9)$$

where α denotes the azimuth of the sectioning plane at the point of interest. By examining the formula, it is clear that $R_\alpha = M$ when $\alpha = 0°$ and that $R_\alpha = N$ when $\alpha = 90°$, as expected. Notice that R_α does not depend on longitude.

Example 4.9. Compute the radius of curvature in the normal section on the GRS 80 ellipsoid at a geodetic latitude $\phi = 41.980\,97°$ and at an azimuth $\alpha = 45°$.

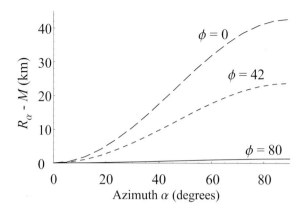

Figure 4.12: The graph of $R_\alpha - M$ (km) for azimuths from $0°$ to $90°$ at $\phi = 0°, 42°,$ and $80°$.

Solution:

$$\frac{1}{R_\alpha} = \frac{\cos^2 \alpha}{M} + \frac{\sin^2 \alpha}{N}$$
$$= \frac{\cos^2 45°}{6\,364\,009.194\,79} + \frac{\sin^2 45°}{6\,387\,710.095\,74}$$
$$= 1.568\,421\,374\,065\,443\,1398 \times 10^{-7}$$

so

$$R_\alpha = 6\,375\,838.$$

rounded to the nearest meter.
□

Example 4.10. What is the length on the GRS 80 reference ellipsoid of 1 second of arc at the equator at an azimuth of $45°$?

Solution: Over such a short distance, the answer can be given by $d = r\,\theta$. Taking r to be R_α with $\phi = 0°$ and $\alpha = 45°$ gives $R_\alpha = 6\,356\,716.465$ m. With $\theta = 1''$ converted to radians, $d = 30.818$ m. At the equator, one meridional arc second corresponds to a distance of 30.715 m, and one arc second along the equator corresponds to a distance of 30.922 m. □

Eq. 4.7, 4.8, and 4.9 show that $N \geq R_\alpha \geq M$ with equality occurring only at the poles. Figure 4.12 shows three graphs of $R_\alpha - M$ as a function of azimuth at $\phi = 0°, 42°,$ and $80°$. At low latitudes the difference can be upwards of 40 km as α approaches $90°$. The graph also shows how N differs from M because $R_\alpha = N$ for $\alpha = 90°$.

It should also be noted that, although the osculating circle for N is normal to that for M and, thus, is locally oriented east-west at some point of interest, the radius of curvature in the prime vertical is not equal to the radius of curvature in the plane of that point's parallel. The latter quantity is not normal to the surface of the ellipsoid, nor does it have the same length as N, even though, in some sense, they are both aligned east-west.

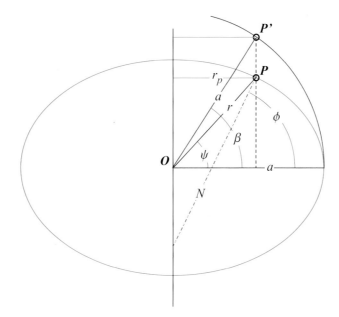

Figure 4.13: Reduced, geocentric, and geodetic latitudes. The reference ellipsoid is in green. The semimajor axis a is purple. Radius of curvature in the prime vertical N is dashed red. Geocentric latitude ψ and associated distance r are shown to the point of interest **P**. The radius of the parallel at **P**'s geodetic latitude ϕ is r_p, shown in light blue. **P'** is a point on a circle of radius a, shown purple. The dashed line perpendicular to a contains **P** and **P'**. Notice how the radius of the parallel for ϕ is $r_p = a \cos \beta$, a very simple expression.

4.5 Types of Latitude

Geodetic latitude is defined as "the *angle* that the normal to an ellipsoid at a point makes with the equatorial plane of that ellipsoid" (NGS 2009). The symbol ϕ stands for geodetic latitude, both in common literature and as adopted by the International Organization for Standardization (ISO) and the International Electrotechnical Commission (IEC) in ISO/IEC18026:2006(E) and ISO19111.

It was noted in section 4.4.2 that geocentric latitude was not generally acceptable for use in positions, because it lead to the problem that "up" in the direction of geocentric latitude was not normal to the ellipsoid. In contrast, the radii of both osculating circles in meridional and prime vertical planes are normal to their ellipsoid. A line segment normal to an ellipsoid, extending between its surface and its minor axis, forms a natural visual representation for geodetic latitude, which is the angle between that line segment and its projection into the equatorial plane (Fig. 4.11 on page 69).

There are occasions that require knowing a parallel-circle's radius, such as when computing distances of due east-west travel. Neither geodetic latitude nor geocentric latitude provide a convenient way to compute the radius of a circle of latitude on an ellipsoid, which motivated the creation of reduced latitude; see angle β in Fig. 4.13. The formula for **reduced latitude** is

$$\beta = \arctan([1 - f] \tan \phi) \qquad (4.10)$$

where f is flattening, and ϕ is geodetic latitude. A parallel of latitude's radius is $r_p = a \cos \beta$, where a is the length of the semimajor axis. The following relationships hold (see Rapp 1989a, p. 25, and Ewing and Mitchell 1970, p. 23):

$$\tan \psi = (1 - \epsilon^2) \tan \phi \quad = \frac{b^2}{a^2} \tan \phi$$

$$\tan \psi = (1 - \epsilon^2)^{1/2} \tan \beta = \frac{b}{a} \tan \beta$$

$$\tan \beta = (1 - \epsilon^2)^{1/2} \tan \phi = \frac{b}{a} \tan \phi$$

$$\tan \beta = \frac{\tan \psi}{(1 - \epsilon^2)^{1/2}} \quad = \frac{a}{b} \tan \phi$$

$$\tan \phi = \frac{\tan \psi}{1 - \epsilon^2} \quad = \frac{a^2}{b^2} \tan \psi$$

$$\tan \phi = \frac{\tan \beta}{(1 - \epsilon^2)^{1/2}} \quad = \frac{a}{b} \tan \beta$$

These equations reveal that $\phi > \beta > \psi$.

4.6 Your Turn

Problem 4.1. Eq. 4.2 on page 56 is only approximately correct because higher-order terms were omitted. The largest truncation error comes from the first omitted term: $5c^4/(24\,a^3)$. Compute this term's contribution to the truncation error by setting $a = 6371$ km and $c = 100$ km. What proportion, in parts per million, of 100 km is this term's part of the error?

Problem 4.2. Convert $267.339\,815\,84°$ east longitude to DMS. Express your answer as both east and west longitude.

Problem 4.3. What is the inverse flattening of the Clarke 1866 reference ellipsoid? What is the inverse flattening of GRS 80?

Problem 4.4. What are the (first) eccentricities squared of the Clarke 1866, GRS 80, and WGS 84 reference ellipsoids?

Problem 4.5. Using GRS 80, what is the radius of curvature in the meridian of a point at 41° N?

Problem 4.6. Using GRS 80, what is the radius of curvature in the prime vertical of a point at 41° N?

Problem 4.7. Using GRS 80, what is the radius of curvature in the normal section of a point at 41° N at an azimuth of 0°? Perform the same computation at azimuths of 45°, 90°, 180°, and 270°.

Problem 4.8. Part of the border between Canada and the United States coincides with the 49th parallel of geodetic latitude. Using the Clarke 1866 reference ellipsoid, how long (in meters) is this part of the border, if it spans from 95° W to 123° 19′ W? How long (in meters) is this part of the border using the GRS 80 reference ellipsoid? How many meters longer is the Clarke 1866 length than the GRS 80 length?

Chapter 5

Geodetic Coordinate Systems

Knowledge of coordinate systems is essential for the proper use of digital spatial data, and coordinate system transformations can solve many difficult problems that arise in geomatics.[1] Weisstein (2009) defines a **coordinate system** as "A system for specifying points using coordinates measured in some specified way." Four types of coordinate systems are commonly used in geomatics: (i) Earth-Centered, Earth-Fixed (ECEF) geocentric Cartesian, (ii) local horizontal Cartesian, (iii) geodetic curvilinear, and (iv) projected planimetric Cartesian. It is possible to transform coordinates between all these systems without losing any information, so in that sense, they are equivalent. However, each system is particularly useful either for representing certain types of observations or for solving certain types of problems, so it's common to collect data in one coordinate system and then transform them into another to reach the final objective.

5.1 Earth-Centered, Earth-Fixed Geocentric Cartesian (XYZ)

Geocentric Cartesian coordinate systems are created as part of the realization of **conventional terrestrial reference frames** (CTRF) (see section 5.4 on page 85) by scientific organizations such as the International Astronomical Union (IAU), the International Union of Geodesy and Geophysics (IUGG), and the International Earth Rotation and Reference Systems Service (IERS); universities; and national geodetic surveys. These reference frames are used in, for example, global navigation satellite systems, astronomy, astrophysics, geodesy, geodynamics, and climate change studies. They provide the means by which all scientifically accurate large-extent positioning is done.

The definition of a **reference system** is given by Kovalevsky et al. (1989, p. 2) as "...a general statement giving the rationale for an ideal case, i.e., for an *ideal reference system*." Reference systems are akin to blueprints – they put forth the conceptual basis upon which an actual house can be built. Reference systems necessarily contain a certain number of arbitrary choices, such as the location of the prime meridian. These choices constitute conventions, so coordinate systems defined with conventions are called **conventional**. Conventional reference systems exist for astronomy and for Earth sciences. The former are called **conventional celestial reference systems** (CCRS), and the latter are called **conventional terrestrial reference systems** (CTRS). One major difference between a celestial reference system and terrestrial references system is that the

---

[1] The mathematics in this and subsequent chapters depends on vector and linear algebra, so readers might want to refer to appendix A for a review, if necessary.

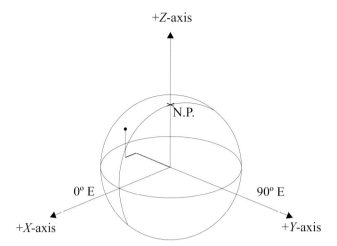

Figure 5.1: The ITRF2000 XYZ coordinates of the southeast corner of the roof of the W. B. Young Building on the University of Connecticut campus ($X = 1\,451\,658$, $Y = -4\,534\,383$, $Z = 4\,230\,142$). The coordinates indicate distances in meters along the axes away from the Earth's center of mass: black = X, red = Y, blue = Z. N.P. = North Pole.

latter is designed so that it rotates with Earth's diurnal rotation; it is fixed to Earth. If this were not so, the coordinates of any stationary place on Earth would constantly change in longitude, which is obviously undesirable.

The International Terrestrial Reference System (ITRS) is an example of a CTRS. According to the International Earth Rotation Service (McCarthy 1996), a coordinate system is a conventional terrestrial reference system if it satisfies these criteria:

- Its origin is at Earth's center of mass including the oceans and atmosphere and, therefore, geocentric (hence *Earth–Centered*).

- Its scale is that of the local Earth frame, in the meaning of a relativistic theory of gravitation.

- Its orientation was initially given by the Bureau International de l'Heure (BIH) orientation of 1984.0.

- Its time evolution in orientation will create no residual global rotation with regards to the crust (hence *Earth–Fixed*).

Reference systems are used to create **reference frames**. According to Kovalevsky et al. (1989, p. 2), "The purpose of a *reference frame* is to provide the means to materialize a *reference system* so that it can be used for the quantitative description of positions and motions on Earth (terrestrial frames), or of celestial bodies, including Earth, in space (celestial frames)." Such frames are called **conventional terrestrial reference frames** (CTRF). A realization of the ITRS is called an **International Terrestrial Reference Frame** (ITRF). ITRFs are realized every few years and have a year designation in their names: e.g., ITRF2005 (Altamimi et al. 2007). The century digits are often omitted as a shorthand: e.g., ITRF00 is actually named ITRF2000.

A geocentric Cartesian coordinate system has three mutually perpendicular axes labeled X, Y, and Z. The Z-axis is set parallel to the Earth's mean rotational axis at some moment in time. The

X-axis is perpendicular to the Z-axis. The X-axis and the Z-axis define the plane of the prime meridian (PM). The Y-axis is set at right angles to the other two axes such that the three form a right-handed coordinate system. A **right-handed coordinate system** is one such that, if one were to place one's right fist at the origin, point the fingers along the positive X-axis, and curl them toward the positive Y-axis, the thumb would point in the direction of the positive Z-axis. The equatorial plane is defined by the X- and Y-axes, and is, therefore, perpendicular to the Z-axis.[2] Figure 5.1 on the facing page illustrates the relationships between the axes and how the axes are used to specify positions. Geocentric Cartesian coordinates will usually be denoted as XYZ coordinates hereafter (note the upper case; lowercase xyz will be used for projected planimetric grid coordinates with some kind of height appended).

ITRS coordinates are reported in meters. For example, the ITRF00 coordinates of the southeast corner of the roof of the W. B. Young Building are (1 451 658, -4 534 383, 4 230 142), which means that the point is 1 451 658 meters from the origin in the X direction, -4 534 383 meters from the origin in the Y direction, and 4 230 142 meters from the origin in the Z direction. Connecticut is in the Northern Hemisphere and, consequently, its Z-value is positive. Likewise, it's in the Western Hemisphere, so its Y-value is negative.

5.1.1 XYZ advantages

1. XYZ positions are inherently three–dimensional. It can be easy to forget that all geospatial positions represent locations with respect to a three–dimensional object, the Earth. XYZ positions keep this fact clearly in view.

2. GPS satellites are the reference points from which user positions are determined using GPS receivers. The positions of the satellites determined from the broadcast ephemerides currently refer to the WGS 84(G1150) frame, so positions determined using the broadcast ephemerides likewise refer to WGS 84(G1150). Positions computed using other ephemerides could refer to other frames. For example, NGS currently computes its precise orbits in IGS2005 and Russian *Global'naya Navigatsionnaya Sputnikovaya Sistema* (GLONASS) orbits refer to PZ 90 (Bazlov et al. 1999). But, in all cases, these positions are computed in XYZ coordinates.

3. If we treat XYZ positions as geometric points, the arithmetic difference of two XYZ positions yields the vector describing their (directed) separation. This is the baseline determined by phase differencing processing used in GNSS positioning calculations.

4. An XYZ vector constitutes the straight-line separation between two points in space. The magnitude of this vector is the geometric representation of the distance measured by an EDM, ignoring very small curvature caused by atmospheric inhomogeneities and other very small effects.

5. The great circle distance d_{gc} between two points, given in XYZ coordinates, on a sphere with a known radius r that is centered at $(X, Y, Z) = (0,0,0)$ has a particularly simple form (see Eq. 6.4 in section 6.1.4 on page 108). The great circle distance is the length of the great circle arc connecting the points, which can be found using $d_{gc} = r\theta$, where θ is the spatial angle separating the points in the plane of the great circle, and r is the radius of the sphere.

---

[2] These definitions are more intuitive than formal. The exact technical definitions for the axes can be found in McCarthy (1996).

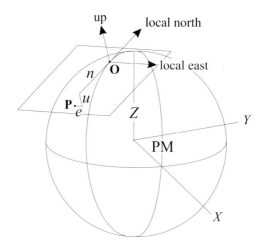

Figure 5.2: Local geodetic East-North-Up (*ENU*) coordinate system.

Either of the (X, Y, Z) points can be interpreted as a vector from the center of the sphere to that point, so the lengths of the vectors both equal r, and the inner product of the vectors can readily be solved for θ.

5.1.2 *XYZ* disadvantages

1. Every *XYZ* position requires three coordinates, as opposed to geographic coordinate system positions, which require only two, if the location's height is neglected.

2. *XYZ* coordinates can be positive or negative and have many digits. This can be awkward and confusing.

3. Consider the *XYZ* coordinates of a ship that is to sail due north. The longitude of the vessel is constant (only the latitude is changing), which makes it easy to steer the desired course. If the ship's position was given in *XYZ*, in general, all three coordinates would be changing simultaneously and continuously. This makes navigation more difficult with *XYZ* coordinates than with other systems.

4. Many people intuitively associate the *Z*-axis with "up." For this reason, *XYZ* coordinates can be confusing. The *Z*-coordinate of an *XYZ* position indicates "up" only at the North Pole.

5. Long distances computed from *XYZ* coordinates do not follow the surface of the Earth; they penetrate it. This can be counterintuitive.

5.2 Geodetic Longitude and Latitude, and Ellipsoid Height (*LBH*)

Geodetic coordinates are geodetic latitude and longitude (section 4.5 on page 71), and ellipsoid height. **Ellipsoid height** is the distance from the surface of a reference ellipsoid to a point of

5.2. GEODETIC LONGITUDE AND LATITUDE, AND ELLIPSOID HEIGHT (LBH)

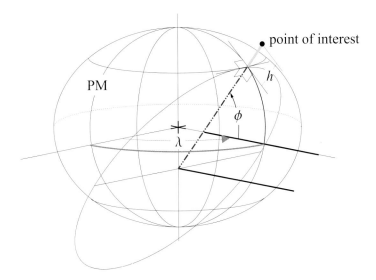

Figure 5.3: Geodetic geographic coordinates.

interest in the direction normal to the ellipsoid (Fig. 5.3). Ellipsoid height is denoted by h (H is reserved for orthometric height [section 10.1.1 on page 191]).

Geodetic positions can be given with either latitude or longitude first. The sequence latitude-longitude-height is traditional, especially with mariners, but it forms a left-handed coordinate system (Y, X, Z). Therefore, geodesists frequently prefer the sequence longitude-latitude-height.

Geodetic coordinates are abbreviated B, L, and H, which are the first letters of the German words for latitude, longitude, and height (*Breite, Länge*, and *Höhe*). This is preferable to the English letters LLH, which do not indicate whether latitude or longitude is first. Therefore, *LBH* and *BLH* are used and distinguish between the left-handed and the right-handed ordering, respectively.

5.2.1 Transforming $LBH \leftrightarrow XYZ$

The formulæ that transform $LBH \to XYZ$ are

$$X = (N + h) \cos \phi \cos \lambda \tag{5.1a}$$
$$Y = (N + h) \cos \phi \sin \lambda \tag{5.1b}$$
$$Z = ((1 - \epsilon^2)N + h) \sin \phi \tag{5.1c}$$

where λ is geodetic longitude, ϕ is geodetic latitude, h is ellipsoid height, ϵ is eccentricity, and N is the radius of curvature in the prime vertical at ϕ.

The inverse problem of transforming XYZ coordinates to geodetic coordinates has received a fair amount of attention (Heiskanen and Moritz 1967; Bowring 1976; Taff 1985; Borkowski 1989; Laskowski 1991; Lin and Wang 1995; Vaníček and Krakiwsky 1996; Fukushima 1999; Gerdan and Deakin 1999; You 2000; Jones 2002; Pollard 2002; Vermeille 2002; Fukushima 2006; Feltens 2008); Featherstone and Claessens (2008) provides a review. The exact solution can be found in Heikkinen (1982). Bowring (1976) developed one of the most widely used procedures. Although it is iterative, the procedure is always adequate without iteration for determining positions on or near the Earth's surface, with an accuracy of about $1\,\mu$m. Bowring amended his original height formula to make it applicable for places in outer space (Bowring 1985). Fukushima (1999) introduced a procedure that is accurate at all heights, thus improving a shortcoming of Bowring's procedure, which loses accuracy near the Earth's geometric center (**geocenter**). We present Bowring's noniterative procedure due to its widespread use and improved accuracy for height (Bowring 1985). Nevertheless, this formula can be inadequate for space geodesy purposes, so more modern approaches are recommended, such as Fukushima (2006). Let $p = \sqrt{(X^2 + Y^2)}$. Then

$$\tan \lambda = Y/X \tag{5.2a}$$

$$\tan k = \frac{Z}{(1-f)p} \tag{5.2b}$$

$$\tan(90° - k) = \frac{(1-f)p}{Z} \tag{5.2c}$$

$$\tan \phi = \frac{Z + (\epsilon')^2 b \sin^3 k}{p - \epsilon^2 a \cos^3 k} \tag{5.2d}$$

$$\tan(90° - \phi) = \frac{p - \epsilon^2 a \cos^3 k}{Z + (\epsilon')^2 b \sin^3 k} \tag{5.2e}$$

$$h = p \cos \phi + Z \sin \phi - (a^2/N) \tag{5.2f}$$

where a and b are the lengths of the semimajor and semiminor axes, respectively; f is flattening; ϵ^2 is eccentricity squared; $(\epsilon')^2$ is second eccentricity squared; and N is the radius of curvature in the prime vertical at ϕ. k is just an intermediate result. Its only purpose is its use in other equations. Use Eq. 5.2c and Eq. 5.2e when $\phi \approx \pm 90°$.

Note that in the above equations, λ is not given explicitly. $\tan \lambda$ occurs in Eq. 5.2a, so $\lambda = \arctan(Y/X)$. Angles of interest are often left embedded within a trigonometric function, requiring their extraction using an inverse trigonometric function. This embedding is a common practice because the term often appears somewhere else in the equations.

Both λ and ϕ are embedded within tangent functions. Expressed in radians, the domain of geodetic latitude is $\pi/2 \leq \phi \leq -\pi/2$, so a single-argument tangent function (as on a hand calculator) will give the correct result. However, longitude's domain spans (at least) 2π radians, so a single-argument tangent function will not suffice and the quadrant correction term given in Eq. 3.8 on page 33 must be used (or a two-argument tangent function such as available in programming languages and spreadsheets [e.g., ATAN2]).

These equations show that the choice of the reference ellipsoid has an impact on the values of the geodetic coordinates. Therefore, one cannot give geodetic positions without indicating which ellipsoid they refer to. This is not true for XYZ positions. They can be, and are, produced without any notion of a reference ellipsoid whatsoever.

Example 5.1. Convert NAD 83 (41° 21′ 12.994 87″N, 72° 01′ 25.040 41″W, 635.478) to XYZ coordinates.
Solution: In decimal degrees, the geodetic latitude and longitude are 41.353 609 686° and $-72.023\,622\,336°$, respectively. The formula for the radius of curvature in the prime vertical is Eq. 4.8:

$$N = a/\sqrt{(1 - \epsilon^2 \sin^2 \phi)}$$

where a is the length of the semimajor axis of the ellipsoid, ϵ^2 is eccentricity squared, and ϕ is

5.2. GEODETIC LONGITUDE AND LATITUDE, AND ELLIPSOID HEIGHT (LBH)

geodetic latitude. So

$$N = 6\,378\,137/\sqrt{(1 - 0.006\,694\,380\,022\,90\,\sin^2 41.353\,609\,6861°)}$$
$$= 6\,387\,476.887$$

in meters. From Eq. 5.1a

$$X = (N + h)\cos\phi\cos\lambda$$
$$= (6\,387\,476.887 + 635.478)\cos(41.353\,609\,6861°)\cos(-72.023\,622\,336\,11°)$$
$$= 1\,479\,921.839$$

From Eq. 5.1b,

$$Y = (N + h)\cos\phi\sin\lambda$$
$$= (6\,387\,476.887 + 635.478)\cos(41.353\,609\,6861°)\sin(-72.023\,622\,336\,11°)$$
$$= -4\,561\,128.808$$

From Eq. 5.1c,

$$Z = ((1 - \epsilon^2)N + h)\sin\phi$$
$$= ((1 - 0.006\,694\,379\,99)\,6\,387\,476.887 + 635.478)\sin(41.353\,609\,6861°)$$
$$= 4\,192\,401.531$$

□

Example 5.2. Convert the XYZ coordinates from Example 5.1 back into geodetic coordinates.
Solution: From Eq. 5.2a

$$\lambda = \tan^{-1}(Y/X)$$
$$= \tan^{-1}(-4\,561\,128.808\,\text{m}/1\,479\,921.839)$$
$$= \tan^{-1}(-0.324\,463\,943\,364)$$
$$= -72.023\,622\,336°$$

From Eq. 5.2b

$$k = \tan^{-1}\left(\frac{Z}{(1-f)\sqrt{(X^2+Y^2)}}\right)$$
$$= \tan^{-1}\left(\frac{4\,192\,401.531}{(1 - 1/298.257\,222\,101)\sqrt{(1\,479\,921.839^2 + (-4\,561\,128.808)^2)}}\right)$$
$$= \tan^{-1} 0.877\,230\,152$$
$$= 41.258\,215\,23°$$

Compute $(\epsilon')^2 = (a^2 - b^2)/b^2 = (6\,378\,137^2 - 6\,356\,752.314^2)/6\,356\,752.314^2 = 0.006\,739\,496\,775\,47$.
$p = \sqrt{X^2 + Y^2} = \sqrt{(1\,479\,921.839^2 + (-4\,561\,128.808)^2)} = 4\,795\,212.681$. From Eq. 5.2d

$$\begin{aligned}\phi &= \tan^{-1}\left(\frac{Z + (\epsilon')^2 b \sin^3 k}{p - \epsilon^2 a \cos^3 k}\right) \\ &= \tan^{-1}\left(\frac{4\,192\,401.531 + 0.006\,739\,496\,775\,47 \times 6\,356\,752.314 \sin^3 41.258\,215\,228°}{4\,795\,212.681 - 0.006\,739\,496\,775\,47 \times 6\,378\,137 \cos^3 41.258\,215\,228°}\right) \\ &= \tan^{-1}(0.880\,180\,642\,4) \\ &= 41.353\,609\,686°\end{aligned}$$

And, finally, from Eq. 5.2f

$$\begin{aligned}h &= \frac{p}{\cos\phi} - N(\phi) \\ &= \frac{4\,795\,212.681}{\cos 41.353\,609\,6861°} - 6\,387\,476.887\,01 \\ &= 635.478\end{aligned}$$

□

5.2.2 Geodetic coordinate advantages

1. Latitude and longitude are traditional and well known.

2. Geodetic coordinates are a natural way to represent positions on the Earth.

3. Geodetic coordinates are extremely convenient for global navigation.

4. Geodetic coordinates represent positions without any of the distortions caused by cartographic projections. Therefore, their interpretation is more straightforward.

5. Survey control networks have control coordinates published using geodetic coordinates.

6. Geodetic coordinates are a unified scheme for the whole Earth, a scheme that does not break it up into zones.

5.2.3 Geodetic coordinate disadvantages

1. Spherical coordinates are angles, not distances. This can be confusing because most people are used to Cartesian coordinates, which are distances.

2. In general, the spherical coordinate formulæ are more complicated than their Cartesian equivalents. The ellipsoidal formulæ are usually much more complicated, sometimes not having closed-form algebraic expressions.

3. Calculating the shortest distance between two points on an ellipsoid (the length of a geodesic) involves elliptic integrals, which have no closed-form algebraic expression.

4. Geodetic coordinates depend on the ellipsoid they refer to.

5.3 Local Horizontal Coordinate Systems: East-North-Up (*ENU*)

In a local horizontal coordinate system, the east, north, and up axes are centered at a place that is usually in the middle of a region of interest on or near Earth's surface, rather than at the geocenter. The *ENU* coordinate system is very similar to the *XYZ* coordinate system. Both are Cartesian, which means that they have three mutually perpendicular axes, the axes have the same scale, and the coordinates are distances from their respective origins along the coordinate axes. Indeed, any particular *ENU* coordinate system is simply an *XYZ* coordinate system translated from its origin to the local *ENU* origin, and then rotated so that the axes align with the local cardinal directions. In some applications, the destinations of the *X*- and *Y*-axes are reversed for convenience. In an *ENU* coordinate system, an origin **O** is chosen in either *XYZ* or *LBH* coordinates, and the *ENU* north axis is oriented along the meridian passing through **O**, positive to the north. The east axis is oriented along the parallel passing through **O**, positive to the east. The up axis passes through **O** and is perpendicular to the reference ellipsoid at **O**. In Fig. 5.2 on page 76, **O** defines the origin of a local *ENU* coordinate system. In this system, **P** has *ENU* coordinates of (e, n, u) (using lowercase letters helps avoid confusion between local north and the radius of curvature in the prime vertical N).

ENU coordinate systems are very useful with optomechanical surveying, GNSS positioning, and photogrammetry (Wolf and Dewitt 2000). A traverse can be thought of as a series of observations in a sequence of *ENU* coordinate systems, one for each total station setup. In GNSS positioning, in good observation conditions the east and north variances of the baseline vectors are expected to be similar in magnitude, whereas the variance in the up direction is expected to be roughly twice that of the other two. These variances have no obvious pattern when the vector is in an *XYZ* coordinate system, so presenting them in *ENU* provides a convenient way of rapidly judging a vector's quality. In photogrammetry, constructing an *ENU* coordinate system at an aerial camera's film platen provides a natural coordinate system for aerotriangulation and orthorectification (Wolf and Dewitt 2000). This natural relationship between the coordinate system and the physical reality of the measuring device simplifies the mathematics used in photogrammetry.

5.3.1 Transforming $XYZ \leftrightarrow ENU$

Transforming *XYZ* to local horizontal *ENU* involves translating the coordinate system origin from the geocenter to the local origin and then rotating the axes to align the *Y*-axis with north, the *X*-axis with east, and the *Z*-axis with the direction normal to the reference ellipsoid (Wolf and Dewitt 2000, p. 573–576). We need the coordinates of the *ENU* coordinate system's local origin both in geodetic coordinates (ϕ_0, λ_0) and in *XYZ* coordinates (X_0, Y_0, Z_0). Let (X, Y, Z) be the *XYZ* coordinates of the point for which *ENU* coordinates (e, n, u) are needed. Then

$$\begin{bmatrix} e \\ n \\ u \end{bmatrix} = \begin{bmatrix} -\sin\lambda_0 & \cos\lambda_0 & 0 \\ -\sin\phi_0\cos\lambda_0 & -\sin\phi_0\sin\lambda_0 & \cos\phi_0 \\ \cos\phi_0\cos\lambda_0 & \cos\phi_0\sin\lambda_0 & \sin\phi_0 \end{bmatrix} \cdot \begin{bmatrix} X - X_0 \\ Y - Y_0 \\ Z - Z_0 \end{bmatrix} = \mathbf{M} \cdot \begin{bmatrix} X - X_0 \\ Y - Y_0 \\ Z - Z_0 \end{bmatrix} \quad (5.3)$$

Letting $\mathbf{E} = (e, n, u)^T$, $\mathbf{X} = (X, Y, Z)^T$, and $\mathbf{X}_0 = (X_0, Y_0, Z_0)^T$ allows Eq. 5.3 to be written in linear algebra notation as

$$\mathbf{E} = \mathbf{M}.(\mathbf{X} - \mathbf{X}_0) \quad (5.4)$$

The inverse problem of finding the XYZ coordinates of a point whose coordinates are given in ENU is found from Eq. 5.3 by inverting $\mathbf{M}$ and solving for (X, Y, Z) as

$$\begin{bmatrix} X \\ Y \\ Z \end{bmatrix} = \mathbf{M}^{-1} \begin{bmatrix} e \\ n \\ u \end{bmatrix} + \begin{bmatrix} X_0 \\ Y_0 \\ Z_0 \end{bmatrix} \quad (5.5)$$

$\mathbf{M}$ is an **orthogonal matrix**, meaning its inverse equals its transpose: $\mathbf{M}^{-1} = \mathbf{M}^T$. Equation 5.5 can, therefore, be written in linear algebra notation as

$$\mathbf{X} = \mathbf{M}^{-1}.\mathbf{E} + \mathbf{T} = \mathbf{M}^T.\mathbf{E} + \mathbf{T}$$

5.3.2 *ENU* advantages

1. *ENU* coordinates are very intuitive because they reflect human-scale experience of spatial locations and relationships.

2. The origin of an *ENU* coordinate system can be placed anywhere that is convenient.

3. *ENU* coordinate systems fully capture the three-dimensional shape of Earth.

4. *ENU* coordinate systems are geometric representations of the coordinate system of a total station, once the observations are reduced from natural to geometric values (e.g., correcting for the deflection of the vertical).

5. Treating *ENU* positions as geometric points, the arithmetic difference of two *ENU* positions yields the vector describing their (directed) separation. The magnitude of this vector is the geometric representation of the distance measured by an EDM (slope distance) in the absence of environmental factors that cause the EDM laser beam to not follow a straight path.

5.3.3 *ENU* disadvantages

1. *ENU* coordinates will usually take positive and negative values if the origin is placed in the middle of the area of interest.

2. The up coordinate can be counterintuitive. For example, consider any *ENU* coordinate system whose origin is on a reference ellipsiod. With the exception of the *ENU* origin, any point on the reference ellipsoid's surface will have a negative up coordinate because the projection plane is tangent to the ellipsoid at the local origin. Any other point on the ellipsoid's surface is below this tangent plane (e.g, see **P** in Fig. 5.2 on page 76).

3. The U-axis (up) of an *ENU* coordinate system is normal to the reference ellipsoid. Consequently, the up-axes of two different *ENU* coordinate systems are not parallel (see Fig. 6.2 on page 105). This means that the horizontal distance between two stations can differ between their respective *ENU* coordinate systems (see section 6.1.2 on page 105).

5.3. LOCAL HORIZONTAL COORDINATE SYSTEMS: EAST-NORTH-UP (ENU)

5.3.4 *ENU* application

Chapter 3 described traversing using plane surveying. A single planar surface was used for all computations, but in reality, each setup constitutes its own *ENU* coordinate system with its own unique tangent plane. Example 5.3 shows how to create the *ENU* coordinate system of the forward stations, which removes the flat-Earth assumption.

Example 5.3. Suppose we know that

- the NAD 83(CORS96) coordinates (ellipsoid height, in meters, referring to GRS 80) of an observation station are $(41°0'0.00000''$ N, $105°0'0.00000''$ W, $150.000)$,

- the geodetic azimuth to a forward station **F** is $\alpha = 291°18'45''$,

- the zenith angle to **F** is $\zeta = 91°55'20''$, and

- the slope distance to **F** is $d_s = 151.340$ m.

(i) Compute the coordinates of **F** in the local *ENU* system, and (ii) compute the *XYZ* coordinates of **F** to serve as its own *ENU* origin.

Solution: Computation of the *ENU* coordinates follows the same steps as a plane-surveying traverse. Convert angles from DMS to radians to use them in computations: $\alpha = 5.084\,362\,277$ and $\zeta = 1.604\,345\,434$. Slope distance is reduced to a horizontal distance by Eq. 3.9: $d_h = d_s \sin \zeta = 151.255$ m. Slope distance is reduced to a vertical distance by Eq. 3.10: $d_v = d_s \cos \zeta = $ -5.076 m.

(i) Compute the *ENU* coordinates of **F** using Eq. 3.17a and 3.17b on page 43. The coordinates of the origin of any *ENU* system are always $\mathbf{O} = (0,0,0)$, so

$$\begin{aligned}
e_\mathbf{F} &= e_\mathbf{O} + d_h \sin(\alpha) \\
&= 0 + 151.255 \sin(5.084\,362\,277) \\
&= -140.911 \\
n_\mathbf{F} &= n_\mathbf{O} + d_h \cos(\alpha) \\
&= 0 + 151.255 \cos(5.084\,362\,277) \\
&= 54.974
\end{aligned}$$

The up coordinate equals the vertical distance, so $u_\mathbf{F} = $ -5.076.

(ii) The radius of curvature in the prime vertical at the observation station is computed with Eq. 4.8:

$$\begin{aligned}
N &= \frac{a}{(1 - \epsilon^2 \sin^2 \phi)^{1/2}} \\
&= 6\,378\,137/(1 - 0.006\,694\,380\,022\,90 \sin(41°)^2)^{1/2} \\
&= 6\,387\,345.731
\end{aligned}$$

Use Eq. 5.1 to transform (41°0′0.000 00″ N, 105°0′0.000 00″ W, 150.000) to XYZ:

$$\begin{aligned}
X_0 &= (N+h)\cos\phi\cos\lambda \\
&= (6\,387\,345.731 + 150)\cos 41° \cos(-105°) \\
&= -1\,247\,690.063 \\
Y_0 &= (N+h)\cos\phi\sin\lambda \\
&= (6\,387\,345.731 + 150)\cos 41° \sin(-105°) \\
&= -4\,656\,442.709 \\
Z_0 &= ((1-\epsilon^2)N + h)\sin\phi \\
&= ((1 - 0.006\,694\,380\,022\,90)6\,387\,345.731 + 150)\sin 41° \\
&= 4\,162\,521.609
\end{aligned}$$

Computing $\mathbf{M}$ from Eq. 5.3 gives

$$\mathbf{M} = \begin{pmatrix} 0.965\,926 & -0.258\,819 & 0. \\ 0.169\,801 & 0.633\,704 & 0.754\,71 \\ -0.195\,333 & -0.728\,993 & 0.656\,059 \end{pmatrix}$$

Since $\mathbf{M}^{-1} = \mathbf{M}^T$

$$\mathbf{M}^{-1} = \begin{pmatrix} 0.965\,926 & 0.169\,801 & -0.195\,333 \\ -0.258\,819 & 0.633\,704 & -0.728\,993 \\ 0. & 0.754\,71 & 0.656\,059 \end{pmatrix}$$

Use Eq. 5.5 to transform the forward station's ENU coordinates to XYZ:

$$\begin{aligned}
\begin{bmatrix} X \\ Y \\ Z \end{bmatrix} &= \mathbf{M}^{-1} \begin{bmatrix} e \\ n \\ u \end{bmatrix} + \begin{bmatrix} X_0 \\ Y_0 \\ Z_0 \end{bmatrix} \\
&= \begin{pmatrix} 0.965\,926 & 0.169\,801 & -0.195\,333 \\ -0.258\,819 & 0.633\,704 & -0.728\,993 \\ 0. & 0.754\,71 & 0.656\,059 \end{pmatrix} \begin{bmatrix} -140.911 \\ 54.974 \\ -5.076 \end{bmatrix} + \begin{bmatrix} -1\,247\,690.063 \\ -4\,656\,442.709 \\ 4\,162\,521.609 \end{bmatrix} \\
&= \begin{bmatrix} -1\,247\,815.847 \\ -4\,656\,367.701 \\ 4\,162\,559.769 \end{bmatrix}
\end{aligned}$$

□

If the observations in this example were perfect, then $\mathbf{F}$'s geodetic coordinates (use Eq. 5.2 to transform XYZ coordinates to LBH) would be exact because of the application of the coordinate system transformation, even though the forward coordinates were computed using plane surveying techniques. Zenith angle is the most error prone of the three observations; it can be so erroneous that it corrupts the vertical distance well beyond acceptable tolerances, which affects all three XYZ coordinates. The sequence of transformations $ENU \to XYZ \to LBH$ illustrates how to do geodetic surveying using a total station and coordinate system transformations. The next chapters will go through the details of how this is done using reductions, which is the more complicated, and widely used, process.

5.4 Reference Frames and Geodetic Datums

5.4.1 Geodetic datums

Geodetic datum is defined by NGS as "A set of constants specifying the coordinate system used for geodetic control, i.e., for calculating coordinates of points on Earth." Although somewhat vague concerning exactly what a geodetic datum *is*, this definition makes a strong connection between datums and terrestrial geodetic positioning. In fact, it asserts that datums are *necessary* for geodetic positioning – indeed, a datum is a realization of a coordinate system suitable for terrestrial geodetic surveying. Traditionally, there were separate geodetic datums for horizontal coordinates (geodetic longitude and latitude), and vertical coordinates (informally meaning heights referred to the mean ocean surface level). Modern geocentric datums are used three-dimensionally as ECEF coordinate systems, so their three coordinates are geometrical, either *XYZ* or *LBH* (ellipsoid heights rather than elevations).

Three horizontal geodetic datums of continental extent (or greater) have been created for North America (Schwarz 1989, p. 13–20) and their defining parameters have changed in character with each new datum. Examining this evolution provides insight into what a datum is. For example, the constants specifying the North American Datum of 1927 (NAD 27) were (i) the coordinates of station MEADES RANCH in Kansas, (ii) the Clarke 1866 reference ellipsoid, and (iii) controlling azimuths, called **Laplace azimuths**, between various stations (Schwarz 1989, p. 5). The coordinates of MEADES RANCH defined the origin of the coordinate system. Controlling azimuths defined the orientation of the graticule and ensured its directional consistency throughout the datum's extent. These two constants, an origin and orientation, are clearly necessary for any coordinate system. The reference ellipsoid provided a shape and a size of an Earth model upon which to perform the computations to determine the geodetic coordinates of the observation stations forming the triangulation network from which the datum was realized. The Clarke 1866 ellipsoid was custom fit to North America, so it was the best choice at the time. Surveying in NAD 27 is straightforward: find a marker with NAD 27 coordinates to use as a POB and another to use as an azimuth mark, and perform a traverse.

The North American Datum of 1983 (NAD 83) was defined differently (Schwarz 1989, p. ix):

> The fundamental task of NAD 83 was a simultaneous least squares adjustment involving 1,785,772 observations and 266,436 stations in the United States, Canada, Mexico, and Central America. Greenland, Hawaii, and the Caribbean islands were connected to the datum through Doppler satellite and Very Long Baseline Interferometry (VLBI) observations.
>
> The computations were performed with respect to the ellipsoid of the Geodetic Reference System of 1980 (GRS 80), recommended by the International Association of Geodesy (IAG).
>
> The parameters of the ellipsoid are
>
> $$a = 6\,378\,137 \text{ meters (exactly)}$$
> $$1/f = 298.257\,222\,101 \text{ (to 12 significant digits)}$$
>
> The ellipsoid is positioned in such a way as to be geocentric, and the orientation is that of the Bureau International de l'Heure (BIH) Terrestrial System of 1984 (BTS-84). In

these respects, NAD 83 is similar to other modern global reference systems, such as the World Geodetic System 1984 (WGS 84) of the U.S. Defense Mapping Agency (DMA).[3]

The conceptual basis of NAD 83 is very different than that for NAD 27. The GRS 80 reference ellipsoid was placed as geocentrically as was possible at the time the datum was realized. Therefore, the origin of the datum is not on Earth's surface at a survey marker but essentially at Earth's center of mass. It was possible to occupy the origin of NAD 27 (it still is, actually). It is not possible to do so with NAD 83. NAD 27 was oriented by observing celestial bodies. NAD 83 was oriented by placing the GRS 80 semiminor axis parallel to the reference pole of the rotating Earth using Earth rotation parameters (see section 7.1.2 on page 136).

Note the careful choice words in the NAD 83 definition, identifying WGS 84 as a *reference system*, not as a *datum*. The differences between a CTRS and a geodetic datum can be summarized as follows:

1. Not all CTRS have a reference ellipsoid associated with them, notably the BTS mentioned above. This system was the predecessor of the International Terrestrial Reference System (ITRS) and was used for scientific purposes other than terrestrial geodetic surveying, such as astronomy and astrophysics. Positioning celestial bodies does not require a reference ellipsoid, whereas geodetic coordinates cannot be computed without one because basic geodetic positioning computations, such as the transformation from XYZ to LBH (Eq. 5.2), depend upon reference ellipsoid parameters. Terrestrial geodetic surveying requires geodetic coordinates, and those, in turn, require a reference ellipsoid to be associated with the reference frame.

2. The definition of a geodetic datum stipulates that it is used "...for calculating coordinates of points on Earth," i.e., for terrestrial geodetic surveying. Therefore, one could draw a distinction between CTRFs and geodetic datums that CTRFs are not intended to be used for terrestrial positioning (at least directly) whereas geodetic datums are.

3. A CTRF used for terrestrial geodetic positioning is a datum (e.g., the European Terrestrial Reference System).

4. A datum can be a CTRS if it meets the defining criteria (e.g., NAD 83 [Snay and Soler 2000b]).

5.4.2 International Terrestrial Reference System (ITRS)

The realizations of the International Terrestrial Reference System are considered the most accurate CTRFs. The ITRS fulfills the following conditions (McCarthy and Petit 2004):

1. It is geocentric, the center of mass being defined for the whole Earth, including oceans and atmosphere.

2. The unit of length is the metre (SI). This scale is consistent with the TCG[4] time coordinate for a geocentric local frame, in agreement with IAU and IUGG (1991) resolutions. This is obtained by appropriate relativistic modelling.

3. Its orientation was initially given by the BIH orientation at 1984.0.

---

[3]The Defense Mapping Agency is now the National Geospatial-Intelligence Agency (NGA).
[4]Geocentric Coordinate Time

5.4. REFERENCE FRAMES AND GEODETIC DATUMS

4. The time evolution of the orientation is ensured by using a no-net-rotation condition with regards to horizontal tectonic motions over the whole Earth.

The definition of the ITRS is very similar to that of NAD 83. It is realized as an ECEF coordinate system whose Z-axis is "...oriented along Earth's mean rotational axis, the X-axis is in the direction of the adopted origin of longitude and the Y-axis is orthogonal to the X and Z axes and in the plane of the 90° E meridian" (McCarthy and Petit 2004, p. 83).

5.4.3 World Geodetic System 1984 (WGS 84)

WGS 84 (NIMA 1997) is a conventional terrestrial reference system, and its definition follows that of the ITRS (NIMA 1997, p. 2-1). WGS 84 was created by the U.S. Department of Defense (DoD) to support global mapping and navigation applications. Realizations of WGS 84 are the frames in which NGA computes the GPS broadcast orbits and the NGA's precise orbits by using the coordinates of permanent DoD GPS monitor stations. Therefore, coordinates determined using the GPS broadcast ephemerides refer to a WGS 84 reference frame.

WGS 84 realizations are denoted by the letter "G" (indicating the use of GPS techniques) followed by the GPS week number in which the realization was created. There are three WGS 84 reference frames: WGS 84(G730), WGS 84(G873), and WGS 84(G1150). Each successive frame has been created to be closer to ITRS frames. WGS 84(G873) and ITRF94 reveal systematic differences no larger than 2 cm (Malys and Slater 1994), and WGS 84(G1150) is practically identical with ITRF00 (Snay and Soler 2000a).

5.4.4 Geocentric datums

Modern geocentric datums include the following (Soler and Marshall 2003):

- European Terrestrial Reference System of 1989 (ETRS89) (Adam et al. 2002)

- Geocentric Datum of Australia of 1994 (GDA94) (Featherstone 1996)

- North American Datum of 1983(NAD 83) (Schwarz 1989)

- Sistema de Referencia Geocentrico para Las Americas[5] (SIRGAS) (Hoyer et al. 1998)

These datums are conceptually akin to conventional terrestrial reference systems. They are blueprints for creating realizations of the datum. The realizations of these datums are transformations of an ITRS frame and, like the ITRS, these datums can have more than one realization. The realizations of NAD 83 are NAD83(HARN), NAD83(CORS93)[6], NAD83(CORS94), NAD83(CORS96), NAD83(CSRS98), NAD83(PACP00), and NAD83(MARP00). A realization of ETRS89 is called the European Terrestrial Reference Frame of nn (e.g., ETRF89) for nn (year) = 1989-1997 and 2000.

---

[5]South American Geocentric Reference System
[6]CORS: Continuously Operating Reference System

5.4.5 Datum issues when surveying with GNSS

One can still conduct a traditional terrestrial survey in a modern geocentric datum as in an older regional datum: find a marker with coordinates published in the geocentric datum as a POB and another for orientation. However, recall that GPS orbits are computed in WGS 84 and ITRS reference frames – they are not computed in the frames of any geocentric datum. Positions determined using GPS refer to the frame of the orbits, so how is it possible to use GPS to survey in a modern geocentric datum? Since all modern geocentric datums are commensurate with an ITRS frame, equations can transform positions between the frames with very high accuracy, or even exactly. Suppose positions are computed using broadcast ephemerides that refer to WGS 84(G1150), which is practically identical to ITRF00. One applies a frame transformation (section 5.5) from ITRF00 to, say, NAD 83(CORS96) to refer the positions to the desired frame. Surveying using other constellations, such as GLONASS, requires that there be high-accuracy transformations from the reference frame of those orbits to the desired frame, among other criteria (Bazlov et al. 1999).

Frame transformation equations are coded into GNSS processing software, which removes the onus of the transformation from the surveyor. Nevertheless, surveyors need to verify that the transformation is the correct one. Some GNSS processing software packages apply no transformation unless explicitly directed otherwise; the positions will refer to whichever frame was referenced by the orbits, which will be some realization of WGS 84 if International GNSS Service (IGS) precise orbits are not used. WGS 84 and NAD 83 frames are very similar; they place the centers of their reference ellipsoids apart by roughly 2 m. Mistakenly leaving the coordinates in WGS 84 introduces a subtle, and possibly difficult to detect, blunder.

Modern geocentric datums are designed so that the GNSS surveyor seems to work directly in the datum. This is not true when the GNSS surveyor works in regional datums, such as NAD 27, or in arbitrary local datums. Legacy datums often lack the spatial self-consistency of modern datums, which makes it impossible to develop a survey-accuracy Helmert transformation between a legacy datum and an orbit's frame. For example, the NAD 27 graticule is somewhat distorted compared to the NAD 83 frames, and the distortion is unpredictable. Transformations between NAD 83 and NAD 27 are best done using NADCON, a computer program that uses spline surface-patch interpolation (Farin 1993), custom fitting individual patches of the Earth from one frame to the other (Dewhurst 1990). NADCON produces NAD 27 positions with accuracies usually acceptable for, say, resource location studies or datum transformations in a GIS that do not require survey-grade accuracy. These accuracies are usually not acceptable for land surveys, with an uncertainty level of 0.15 m in the conterminous United States (NADCON 1990). NADCON is "not appropriate to transform FGCC first- or second-order coordinates between NAD 27 and NAD 83 and retain first- or second-order accuracies in the results... recomputation or readjustment of survey observations is usually more appropriate to maintain first- and second-order FGCC accuracies"[7] (NADCON 1990, p. 32681). However, GNSS can be used to position in local datums using localization. **Localization** is a process whereby a transformation between a local datum and the GNSS frame is computed on an as-needed basis. Localization is typically done by occupying at least three stations with known local coordinates to determine their positions in the GNSS frame. Knowing the stations' positions in both systems allows spline surface-patches to be computed, which are used to interpolate positions in between the stations. Localized positions will attain survey quality if the local datum distortion is sufficiently low in the survey area.

---

[7]FGCC: Federal Geodetic Control Committee

5.5 Frame Transformation Formulæ

Modern geocentric datums are designed to fully support GPS positioning, for which they must be consistent with ITRS frames (including WGS 84). This consistency requires that the spatial relationship between the datum's frame and the orbits' ITRS frame be well known: frames usually do not agree on the location of the geocenter, their axes are (usually slightly) rotated away from each other, and frames usually have slightly different scales. Frame transformation formulæ account for these three discrepancies.

There are various types of mathematical approaches to transform positions between geocentric reference frames (Kovalevsky et al. 1989; Rapp 1989b; NIMA 1997; Soler 1998; Snay and Soler 2000b; Soler and Marshall 2002; Soler and Marshall 2003; Soler and Snay 2004; Jekeli 2005). Helmert's transformation is quite widely used. In its full generality, Helmert's transformation applies a translation, a rotation, and a scale change to transform an XYZ position from one frame to another, also accounting for the time evolution of the frames with respect to each other and for tectonic plate motion.

5.5.1 Frame translation

Translation can be thought of as a sliding motion that moves something without rotation. Geocenter discrepancies between two frames are accounted for by translating one frame's origin to the other's. A point is translated mathematically by vector addition. If frames $\mathcal{F}$ and $\mathcal{G}$ differ only by the location of their geocenters, then adding a translation vector $\mathbf{T}$ to point $\mathbf{P}^\mathcal{F}$ in frame $\mathcal{F}$ transforms it to the equivalent coordinates $\mathbf{P}^\mathcal{G}$ in frame $\mathcal{G}$.

Translation vectors between nongeocentric frames can be quite large. For example, when transforming from NAD 27 to WGS 72, $\mathbf{T} = (-12.62, 156.56, 176.02)^T$ in meters (Hothem et al. 1982). The length of this vector is 235.91 m. NAD 27 was custom fit to North America, and it was neither possible nor desirable to situate the origin of its reference ellipsoid at the geocenter. So it is not surprising that the translation vector from NAD 27 to a geocentric frame is rather long. In contrast, translation vectors between modern geocentric frames are relatively short, often with lengths on the order of centimeters. For example, the length of the translation from GDA 94 to ITRF00 is $|\mathbf{T}| = 8.9$ cm (Dawson and Steed 2004). The NAD 83(CORS96) origin is offset from the ITRF00 origin by $\mathbf{T} = (0.9956, -1.9013, -0.5215)$, in meters. The length of this vector is 2.210 m, which is the largest translation among the modern geocentric frames. NAD 83 is the oldest of the geocentric frames and its initial realization was established with less accurate methods than what are available nowadays. Once NGS published surveying control coordinates in NAD 83, it was imprudent to move the location of the origin of its reference ellipsoid when better knowledge of the geocenter became available, because to do so would entail a substantial shift of the published control coordinates for no compelling reason. The Helmert transformation from ITRF00 to the NAD 83 frames introduces no error (it is exact), so this minor offset from the geocenter does no harm.

The Helmert transformation for many older regional datums consists only of a translation; nothing is known about discrepancies in scale or frame alignment. Translation-only transformations are called **3-parameter transformations**. The transformation from the Australian Geodetic Datum of 1966 (AGD) to WGS 84 is the following(NIMA 1997):

$$\mathbf{P}^{WGS84} = \mathbf{T} + \mathbf{P}^{AGD} \tag{5.6}$$

where $\mathbf{T} = (-133, -48, 148)$, in meters.

Example 5.4. What are the WGS 84 coordinates of $\mathbf{P}^{AGD} = (151°\,1'\,19''\,\text{E},\,34°\,54'\,30''\,\text{S}, 0)$?
Solution: (i) Transform $\mathbf{P}^{AGD}$ to XYZ. AGD uses the Australian National Spheroid with $a = 6\,378\,160.0$ m and $1/f = 298.25$ (Table 4.1), so $\epsilon^2 = 0.006\,694\,541\,85$. Using Eq. 5.1, $\mathbf{P}^{AGD} = (-4\,580\,718.782,\, 2\,536\,840.879,\, -3\,629\,544.493)$. (ii) Using Eq. 5.6

$$\begin{aligned}\mathbf{P}^{WGS84} &= \mathbf{T} + \mathbf{P}^{AGD} \\ &= (-133, -48, 148) + (-4\,580\,718.782,\, 2\,536\,840.879,\, -3\,629\,544.493) \\ &= (-4\,580\,851.782,\, 2\,536\,792.879,\, -3\,629\,396.493)\end{aligned}$$

(iii) Apply Eq. 5.2 using the WGS 84 reference ellipsoid to transform to BLH: $\mathbf{P}^{WGS84} = (151°\,1'\,23''\,\text{E},\,34°\,54'\,24''\,\text{S})$ at an ellipsoid height of 14 m. $\mathbf{P}^{WGS84}$ is $4''$ different in longitude and $6''$ different in latitude from the coordinates of $\mathbf{P}^{AGD}$. One arc second of longitude spans roughly 30 m at the equator, so one arc second of longitude spans roughly $30\,(35/90)$ m at $35°$ S. Therefore, these points are about $\sqrt{(4 \times 30(35/90))^2 + (6 \times 30)^2} \approx 186$ m apart, ignoring the ellipsoid height difference.
□

5.5.2 Frame scale

Modern reference frames are realized by assigning ECEF coordinates to fiducial monitoring stations: "It is of general agreement now that the frame of the Conventional Terrestrial System is to be defined on the basis of an adopted set of tridimensional coordinates of a global network of tracking stations" (Boucher and Altamimi 1986). Altamimi et al. (2002) wrote, "The basic idea of ITRF is to combine station positions (and velocities) computed by various analysis centers, using observations of space geodesy techniques, such as very long baseline interferometry (VLBI), lunar and satellite laser ranging (LLR and SLR), Global Positioning System (GPS), and Doppler orbitography radiopositionning integrated by satellite (DORIS)." These techniques are used to determine interstation baselines and the Earth's orientation and rotation.

The equation relating baselines between stations and station coordinates is the following (Altamimi et al. 2002, p. 2–3):

$$\begin{pmatrix} \Delta X_b^{i,j} \\ \Delta Y_b^{i,j} \\ \Delta Z_b^{i,j} \end{pmatrix} = \begin{pmatrix} X^j - X^i \\ Y^j - Y^i \\ Z^j - Z^i \end{pmatrix} \tag{5.7}$$

where $(\Delta X_b^{i,j}, \Delta Y_b^{i,j}, \Delta Z_b^{i,j})$ are the geometric vector components of the (observed) baseline for a data set b arising from the observations between fiducial stations i and j, and where, for example, X^i is station i's (adopted) X ECEF coordinate; likewise for station j and the other coordinates. The system of equations that arises by taking Eq. 5.7 for many station pairs and for many observation sets can be expressed as

$$\mathbf{M}.\mathbf{X} = \Delta \mathbf{X}$$

where $\mathbf{X}$ is a vector of the station coordinates, $\Delta \mathbf{X}$ is a vector of the baseline components, and $\mathbf{M}$ is a matrix of zeros, ones, and minus ones relating X^i to X^j for all stations and coordinates. $\mathbf{M}$ is singular, so at least one station's coordinates must be chosen in order to solve the system. This choice is the mechanism to ensure, for example, that the prime meridian is located as desired.

5.5. FRAME TRANSFORMATION FORMULÆ

Different frames are typically realized using different stations, although nowadays frame realizations usually incorporate some or all of the ITRS stations. Two different sets of adopted coordinates will not usually yield reference frames with the same exact implied size of the Earth. We should, however, expect the sizes to be very close because the geophysics that went into their creation is consistent and accurate. We will denote the incremental scale difference between two reference frames by s. Incremental scale difference is a small number, often given in parts per billion. For example, incremental scale difference from ITRF00 to NAD 83(CORS96) is 0.062 ppb (Soler and Snay 2004). If $s = 0$, the frames actually have identical scale, which happens by design (Snay 2003; Jekeli 2005). If $s < 0$, then $\mathcal{F}$ is larger than $\mathcal{G}$ (and vice versa).

5.5.3 Frame alignment

A rotation matrix multiplication is used to align two frames' axes. The alignment matrix $\mathbf{A}$ is defined by three very small angles ω_X, ω_Y, and ω_Z; one for each XYZ-axis. These angles have magnitudes on the order of milli arc seconds for modern geocentric frames (Soler and Marshall 2002). The transformation between XYZ and ENU also involves a rotation matrix (Eq. 5.3). The $XYZ \leftrightarrow ENU$ rotation can involve large angles, so that rotation matrix requires elements involving sines and cosines of the rotation angles. The very small frame alignment angles allow simplifications to be made (if ω is very small, then $\sin \omega \approx \omega$ and $\cos \omega \approx 1$) leading to

$$\mathbf{A} = \begin{bmatrix} 0 & \omega_Z & -\omega_Y \\ -\omega_Z & 0 & \omega_X \\ \omega_Y & -\omega_X & 0 \end{bmatrix}$$

The rotation angles have the sign convention that a positive angle is a counterclockwise rotation when viewed along the positive coordinate axis towards the origin. However, the sign convention for the angles have been given two different meanings by different organizations. The International Earth Rotation and Reference Systems Service (IERS) interprets the angles to be rotations of the position around the coordinate axis, whereas other organizations, such as some national geodetic surveys, interpret the angles to be rotations of the axes (Soler 1998; Soler and Marshall 2002; Soler and Marshall 2003; Dawson and Steed 2004). See Soler (1998) for full details of the rotation angles.

5.5.4 7-parameter Helmert transformation

The 7-parameter Helmert transformation takes a position $\mathbf{P}^{\mathcal{F}}$ in frame $\mathcal{F}$ to a position $\mathbf{P}^{\mathcal{G}}$ in frame $\mathcal{G}$:

$$\mathbf{P}^{\mathcal{G}} = \mathbf{T} + (1+s)(\mathbf{I} + \mathbf{A}) \cdot \mathbf{P}^{\mathcal{F}} \tag{5.8}$$

where the translation vector $\mathbf{T}$ is the same as for the 3-parameter transformation, s is incremental scale difference, $\mathbf{A}$ is the alignment matrix, and $\mathbf{I}$ is an identity matrix. Equation 5.8 appears in other forms in the literature. In Snay 2003, Eq. 5.8 is

$$\mathbf{P}^{\mathcal{G}} = \mathbf{T} + (1+s)\mathbf{R} \cdot \mathbf{P}^{\mathcal{F}} \tag{5.9}$$

where $\mathbf{R} = \mathbf{I} + \mathbf{A}$. In Jekeli 2000, Eq. 5.8 is

$$\mathbf{P}^{\mathcal{G}} = \mathbf{P}^{\mathcal{F}} + \mathbf{T} + s\,\mathbf{P}^{\mathcal{F}} + \mathbf{A} \cdot \mathbf{P}^{\mathcal{F}} \tag{5.10}$$

Table 5.1: Helmert transformation parameter values from ITRF2000 to NAD 83(CORS96) (Soler and Snay 2004).

t_0 decimal year	T_X m	T_Y m	T_Z m	ω_X mas	ω_Y mas	ω_Z mas	s ppb
1997.00	0.9956	-1.9013	-0.5215	25.915	9.426	11.599	0.62
	$\dot{T}_X$ m/yr	$\dot{T}_Y$ m/yr	$\dot{T}_Z$ m/yr	$\dot{\omega}_X$ mas/yr	$\dot{\omega}_Y$ mas/yr	$\dot{\omega}_Z$ mas/yr	$\dot{s}$ ppb/yr
	0.0007	-0.0007	0.0005	0.067	-0.757	-0.051	-0.18

Table 5.2: Helmert transformation parameter values from ITRF2000 to GDA94 (Dawson and Steed 2004).

t_0 decimal year	T_X m	T_Y m	T_Z m	ω_X mas	ω_Y mas	ω_Z mas	s ppb
2000.00	-0.0761	-0.0101	0.0444	8.765	9.361	9.325	7.935
	$\dot{T}_X$ m/yr	$\dot{T}_Y$ m/yr	$\dot{T}_Z$ m/yr	$\dot{\omega}_X$ mas/yr	$\dot{\omega}_Y$ mas/yr	$\dot{\omega}_Z$ mas/yr	$\dot{s}$ ppb/yr
	0.0110	-0.0045	-0.0174	1.034	0.671	1.039	-0.538

and in Soler and Snay 2004, Eq. 5.8 is

$$\mathbf{P}^{\mathcal{G}} = \mathbf{T} + \{(1+s)\mathbf{I} + \mathbf{A}\} \cdot \mathbf{P}^{\mathcal{F}}$$

These forms are not necessarily exactly mathematically identical, but they differ only by a multiplication of s, which is a number on the order of parts-per-billion, so the difference has no pragmatic consequences. The alternate forms are used either to highlight certain aspects of the transformation or to simplify mathematical developments.

Example 5.5. CORS station CTMA has ITRF00 coordinates $(1\,456\,379.156, -4\,539\,029.382, 4\,223\,420.236)$. Transform $\mathbf{CTMA}^{ITRF00}$ to $\mathbf{CTMA}^{NAD83(CORS96)}$.
Solution: The 7-parameter transformation values occupy the second row of Table 5.1.
From Eq. 5.8

$$\begin{aligned}
\mathbf{CTMA}^{NAD83(CORS96)} &= \mathbf{T} + ((1+s)\mathbf{I} + \mathbf{A}).\mathbf{CTMA}^{ITRF00} \\
&= (0.9956, -1.9013, -0.5215)^T + \\
&\quad \begin{bmatrix} 1.62 \times 10^{-9} & 11.599\,\text{mas} & -9.426\,\text{mas} \\ -11.599\,\text{mas} & 1.62 \times 10^{-9} & 25.915\,\text{mas} \\ 9.426\,\text{mas} & -25.915\,\text{mas} & 1.62 \times 10^{-9} \end{bmatrix} . \\
&\quad (1\,456\,379.156, -4\,539\,029.382, 4\,223\,420.236)^T \\
&= (1\,456\,379.704, -4\,539\,030.837, 4\,223\,420.354)
\end{aligned}$$

□

5.5. FRAME TRANSFORMATION FORMULÆ

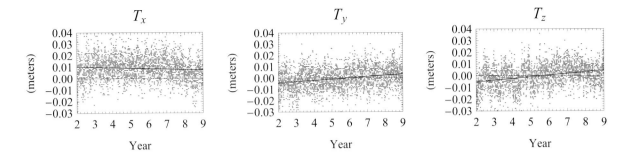

Figure 5.4: WGS 84 to ITRF00 translation vector components from 2002 to 2008 (data courtesy of NGA).

7-parameter Helmert transformations are sometimes applied in sequence. For example, the transformation from ITRF00 to NAD 83(CORS96) involves intermediate transformations (Snay 2003):

$$\text{ITRF00} \to \text{NAD 83(CORS96)} = \text{ITRF00} \to \text{ITRF97} \to \text{ITRF96} \to \text{NAD 83(CORS96)}$$

Helmert transformations are linear, so they can be combined. Table 5.1 on the preceding page gives the combined transformation parameters for ITRF00 → NAD 83(CORS96).

Equation 5.8 can be inverted as follows[8]. For s we have $1/(1+s) \approx 1-s$ because $(1-s)(1+s) = (1-s^2)$, but s^2 is an extremely small number and can safely be ignored. Likewise, $(I+A)^{-1} \approx I$ because $(I+A).(I-A) = I - A.A$ and $A.A$ can safely be ignored, as well. These and similar arguments lead to

$$\mathbf{P}^{\mathcal{F}} = -T + (1-s)(I+A)^{-1} \cdot \mathbf{P}^{\mathcal{G}} \qquad (5.11)$$

The similarity of two reference frames can be judged by the magnitude of their 7-parameter Helmert transformation parameter values. NGA monitors the stability and quality of the GPS constellation by computing a 7-parameter Helmert transformation between WGS 84(G1150) and ITRF00 every day (NGA 2008). The daily 7-parameter transformations are provided to the public in spreadsheet form, one file per year since 2002. Each spreadsheet includes summary statistics. The summary statistics for 2008 are presented in Table 5.1. The translation, rotation, and scale parameters from 2002 to 2008 are plotted in Fig. 5.4, 5.5, and 5.6. Examining the transformation parameters reveals that the two frames have agreed at the centimeter level or better over this time span. The figures reveal long term trends in the parameters, showing that the parameters change over time. The parameter value change rates have an overdot in Tables 5.1 and 5.2. The change rates introduce geocenter displacements on the order of centimeters per year. A 1-mas/year axis rotation results in a positional displacement of $6\,378\,137 \cdot 1$ mas = 3.1 cm at the equator of the WGS 84 reference ellipsoid. The implication is that the evolution of reference frames themselves produces apparent motion in otherwise stationary ground objects on the order of centimeters per year. Such a displacement is readily detectable with survey-grade GNSS positioning. It is, therefore, imperative that survey-grade GNSS positions be transformed from the reference frame of the orbits in which they were computed to the surveying datum using a transformation that accounts for the evolution of the frames.

---

[8]Craig Rollins 2009, personal communication.

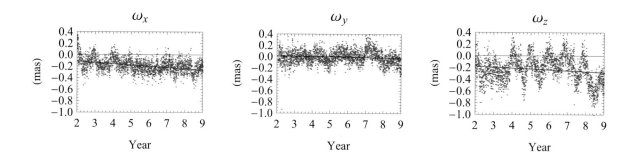

Figure 5.5: WGS 84 to ITRF00 alignment matrix components from 2002 to 2008 (data courtesy of NGA).

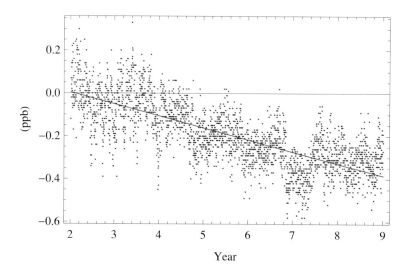

Figure 5.6: WGS 84 to ITRF00 differential scale from 2002 to 2008 (data courtesy of NGA).

5.5.5 14-parameter Helmert transformation

Space geodesy positioning techniques produce positions that are consistent across the entire world. When these positions are examined over time, it is evident that the Earth's crust is in motion. As reported by Snay and Soler (2000b, p. 1),

> ... Earth's center of mass is moving relative to Earth's surface. Also, there are variations of Earth's rotation rate as well as motions of Earth's rotation axis both with respect to space (precession and nutation) and to Earth's surface (polar motion). Moreover, points on Earth's crust are moving relative to one another as a result of plate tectonics, earthquakes, volcanic/magmatic activity, postglacial rebound, people's extraction of underground fluids, solid Earth tides, ocean loading, and several other geophysical phenomena.

Crustal motion can move places each year by distances that are readily observable with GNSS positioning techniques, creating velocity magnitudes on the order of up to 7 cm per year (NIMA 1997, p. 2–6), and episodic displacements of up to several meters due to earthquakes. The plates can be thought of as rigid, thin shells floating on the mantle, being propelled by convection currents in the slowly flowing materials at depth beneath them (Sleep and Fujita 1997). The plates are not sliding across the mantle in a linear (i.e., translations) motion. They rotate about a central point. The rotations are slow and orderly and, therefore, can be predicted. Across geologic time scales, the path of a point on the Earth would need to follow the Earth's curved surface and its plate's nonlinear trajectory. However, for time scales on the order of decades, the distances traveled are so small that the actual nonlinear motion is suitably modeled by a linear translation represented by an XYZ velocity vector. In order for all the velocities over Earth to not suggest an unrealistic expansion or contraction of the Earth as a whole, the ITRS requires (item 4 on page 87 in the ITRS definition above) that the velocities indicate no-net-rotation (NNR) of the tectonic plates, meaning the sum of the velocities over Earth's surface is zero (Argus and Gordon 1991). Thus, ITRF positions are published with XYZ velocity vectors that are nonzero for almost all places on Earth.

Let's stop for a moment to envision the implication for terrestrial surveying of these nonzero velocities and of the time evolution of the reference frames. Terrestrial geodetic surveying, and the land surveying based upon it, needs control markers that are stationary, having coordinates that do not change over time. Geodetic surveys, such as the NGS, publish data sheets that give the authoritative coordinates of survey markers. Here is an example from the NGS data sheet for a bench mark designated Y 88 (NGS assigns a unique, permanent identification [PID] code to all the survey markers in their database which, for Y 88, is LX3030):

```
LX3030 ***********************************************************************
LX3030 CBN            -  This is a Cooperative Base Network Control Station.
LX3030 DESIGNATION -  Y 88
LX3030 PID            -  LX3030
LX3030 STATE/COUNTY-  CT/TOLLAND
LX3030 USGS QUAD   -  COVENTRY (1983)
LX3030
LX3030                           *CURRENT SURVEY CONTROL
LX3030 _____
```

```
LX3030* NAD 83(2007)-  41 48 44.78464(N)    072 15 02.04245(W)     ADJUSTED
LX3030* NAVD 88     -      186.843 (meters)     613.00   (feet)    ADJUSTED
LX3030        _____
LX3030  EPOCH DATE  -        2002.00
LX3030  X           -    1,451,423.337 (meters)                    COMP
LX3030  Y           -   -4,534,399.855 (meters)                    COMP
LX3030  Z           -    4,230,204.319 (meters)                    COMP
```

The two lines indicated with an asterisk give the current survey control coordinates. The reality of tectonic plate motion and the time evolution of reference frames creates a conundrum for control coordinates. LX3030's ITRF00 speed (not shown) is nearly 2 cm per year. Nonzero velocities imply that control coordinates need to be republished frequently, with the extreme being to serve control coordinates in real time. Using, say, ITRF00 coordinates for land surveying would require *reductio ad absurdum* that control coordinates, and all maps showing or based on those coordinates (such as GIS layers), would have to be updated every time they were used. Clearly, this is untenable.

Geocentric datums are created in such a way so that positions referred to that datum have very little or no motion relative to the other positions referred to that datum, as if its positions are fixed in place and time. Geocentric datums have a **reference epoch**: a moment in time at which positions referred to that datum are conceptualized to exist. For example, the reference epochs for NAD 83(CORS96) and NAD 83(2007) are both 2002.00 (NGS documents the reference epoch in a data sheet's EPOCH DATE field). The reference epoch is given in decimal years, so 2002.00 is midnight, January 1, 2002. The 7-Helmert transformation parameters in Eq. 5.8 describe the two frames' geometric relationship at the reference epoch.

Suppose a GNSS survey is controlled by occupying survey markers with GNSS receivers. Control markers will be assigned coordinates in the survey network adjustment – coordinates published by a geodetic authority (such as those in the NGS data sheet above). Published coordinates reflect the location of a marker at the datum's reference epoch. However, that marker is no longer where it was at the reference epoch because of tectonic motion, and the GNSS observations will reflect this. GNSS control coordinates need to be "time warped" forward to the epoch of observation and also transformed into the GNSS orbits' reference frame. The survey network is adjusted in the orbits' frame at the observation epoch. Then, the coordinates are transformed to the datum's reference frame and time warped back to its reference epoch (Snay 1999; Soler and Marshall 2002; Soler and Marshall 2003; Soler and Snay 2004). Capturing time-dependent positioning requires that each of the seven static Helmert parameters have a time-dependent term. Therefore, the time-dependent Helmert transformation has 14 parameters: the 7 static parameters and 7 additional time-dependent parameters. Time-dependent parameters are denoted with an overdot; for example, $\dot{s}$ represents the time-dependent parameter of incremental scale s.

Let δt denote the time differential between the epoch of observation t and the reference epoch of the geocentric datum t_0, given in decimal years: $\delta t = t - t_0$. The observation epoch t is the moment when an observation occurred, such as 0800 UTC 6 March 2008 = $2008 + (31+29+6)/366 + 8/8784 = 2008.18$. Defining δt in this sense (rather than as $\delta t = t_0 - t$) means that the velocities are consistent with time moving forward, but δt will be negative when time warping backwards. The static parameters augmented by their time derivatives are

5.5. FRAME TRANSFORMATION FORMULÆ

$$T_{t,X} = T_X + \delta t\, \dot{T}_X$$
$$T_{t,Y} = T_Y + \delta t\, \dot{T}_Y$$
$$T_{t,Z} = T_Z + \delta t\, \dot{T}_Z$$
$$s_t = s + \delta t\, \dot{s}$$
$$\omega_{t,X} = \omega_X + \delta t\, \dot{\omega}_X$$
$$\omega_{t,Y} = \omega_Y + \delta t\, \dot{\omega}_Y$$
$$\omega_{t,Z} = \omega_Z + \delta t\, \dot{\omega}_Z$$

With this notation, the 14-parameter Helmert transformation has the same form as the seven parameter version. It is different only in the parameters, which reflect their change over δt:

$$\mathbf{P}^{\mathcal{G}}_{t_0} = \mathbf{T}_t + (1 + s_t)(\mathbf{I} + \mathbf{A}_t) \cdot \mathbf{P}^{\mathcal{F}}_t \tag{5.12}$$

where $\mathbf{T}_t = (T_{t,X}, T_{t,Y}, T_{t,Z})^T$ and

$$\mathbf{A}_t = \begin{bmatrix} 0 & \omega_{t,Z} & -\omega_{t,Y} \\ -\omega_{t,Z} & 0 & \omega_{t,X} \\ \omega_{t,Y} & -\omega_{t,X} & 0 \end{bmatrix}$$

Positions have a subscripted epoch to indicate at which epoch these coordinates are valid.

In fact, Eq. 5.12 is a simplification. The full, correct form is derived by taking the time derivative of Eq. 5.8 with respect to t with the understanding that the parameters are actually functions of time. See Soler and Marshall (2002) for the details. The higher-order terms of the fully correct equations were omitted from Eq. 5.12 because omitting them causes no practical degradation in accuracy.

A clever choice of a datum's time-varying alignment parameters allows the motion of a datum's tectonic plate to be built into these parameters so as to cancel the plate's motion with respect to ITRF00. Thus, the datum's plate is effectively fixed; almost all of its positions' velocities are zero or very close to it. Zero velocities imply that positions are immutable; once a position is established, it does not change over time in that reference frame. Plate-fixing Helmert transformations are plate specific; they should not be applied to places outside the datum's tectonic plate. Countries spanning more than one tectonic plate can address this by creating multiple transformations (Snay 2003).

Neither WGS 84 nor ITRS (the two reference systems of GPS) have plate-fixing strategies, because they are world geodetic systems, not surveying datums for a particular country. In general, positions referred to WGS 84 and ITRS frames change by centimeters per year, which is very undesirable for land surveying but very desirable for geodynamics studies. Geomaticians need to understand the time evolution of reference frames in order to use one that is suitable for their purposes.

The following are the ITRF00 and NAD 83(CORS96) positions and velocities for an NGS CORS base station. Note that the ITRF00 velocities are not zero but that the NAD 83(CORS96) velocities are zero.

```
------------------------------------------------------------------------
|                                                                      |
|             Antenna Reference Point(ARP): MANSFIELD CORS ARP         |
|             ------------------------------------------               |
|                             PID = DH5835                             |
|                                                                      |
|                                                                      |
| ITRF00 POSITION (EPOCH 1997.0)                                       |
| Computed in September 2005 using 49 days of data.                    |
|     X =     1456379.156 m      latitude       =  41 43 52.94775 N    |
|     Y =    -4539029.382 m      longitude      = 072 12 38.88087 W    |
|     Z =     4223420.236 m      ellipsoid height =    53.944    m     |
|                                                                      |
| ITRF00 VELOCITY                                                      |
| Predicted with HTDP_2.7 September 2005.                              |
|     VX =    -0.0171 m/yr       northward =    0.0054 m/yr            |
|     VY =    -0.0019 m/yr       eastward  =   -0.0169 m/yr            |
|     VZ =     0.0042 m/yr       upward    =    0.0002 m/yr            |
|                                                                      |
|                                                                      |
| NAD_83 (CORS96) POSITION (EPOCH 2002.0)                              |
| Transformed from ITRF00 (epoch 1997.0) position in Sep. 2005.        |
|     X =     1456379.704 m      latitude       =  41 43 52.91709 N    |
|     Y =    -4539030.837 m      longitude      = 072 12 38.87753 W    |
|     Z =     4223420.354 m      ellipsoid height =    55.182    m     |
|                                                                      |
| NAD_83 (CORS96) VELOCITY                                             |
| Transformed from ITRF00 velocity in Sep. 2005.                       |
|     VX =    -0.0000 m/yr       northward =    0.0000 m/yr            |
|     VY =    -0.0000 m/yr       eastward  =    0.0000 m/yr            |
|     VZ =     0.0000 m/yr       upward    =    0.0000 m/yr            |
|_____|
```

NGS developed a software program named "Horizontal Time Dependent Positioning," or HTDP, that performs all these computations and many more (Snay 1999). Readers wanting to experiment with time-dependent positioning can use the NGS interactive Web site[9] to transform positions and velocities.

5.6 Avoiding Pitfalls

This plethora of datums and coordinate systems, with the added complication of a dynamic Earth, has caused considerable confusion. Some authors have stated that, because the GRS 80 reference ellipsoid is practically identical to the WGS 84 reference ellipsoid, any reference frames based upon them are likewise identical. This is false. Reference ellipsoids can be placed by their datums

---

[9]www.ngs.noaa.gov/TOOLS/Htdp/Htdp.html

5.6. AVOIDING PITFALLS

in different places (disagreement about the geocenter, orientation, and scale), and these can be significantly different. For example, even the difference between NAD 83(CORS96) and ITRF00 shows that it is imperative to transform coordinates determined using GPS positioning from their inherent reference frame to whichever frame is desired for mapping. Fortunately, GNSS processing software packages and most GISs will do almost any frame transformation that is desired, although caution is always needed with transformations to regional datums, which are generally less accurate than the modern geocentric datum transformations.

Most GIS professionals are confronted with data layers referred to different frames. If these data lack metadata that allows them to be automatically transformed into a common frame, then they can misregister with very significant positional shifts. This is readily apparent by looking at the corners of the neat lines of any USGS 7.5-minute topographic map, which show additional crosshairs locating the parallel and meridian intersections for each corner in NAD 27. The offset between NAD 27 and NAD 83 can be dozens of meters. All too often people try to fix this problem by editing one of the data sets, by dragging it with the cursor to try to overlay it one on top of the other. This doesn't work. The shapes of NAD 83 and NAD 27 are different, so simply sliding the layers around will not register them.

Another ad hoc frame transformation approach is called rubber sheeting. **Rubber sheeting** is a collinearity transformation (translation, rotation, and nonuniform rescaling) that forces a set of mutual points to match. Intuitively, it's as if one of the layers was on a rubber sheet that can be stretched and pulled to fit the other by pinning it in place at various mutual points. Rubber sheeting applies the wrong transformation, so the results will not match a Helmert transformation. Rubber sheeting destroys the layer's spatial integrity because it no longer refers to any known frame nor does it have any known relationship to any frame. The proper approach is to determine the frames of the data, and use the datum transformations that are built into the geomatics software to bring the data into a common frame. This approach solves the problem to the best limits of knowledge about the relationships between the frames.

The mismatch between regional and geocentric datums is often obvious, even egregious, and can be magnified by a projection. For example, Universal Transverse Mercator NAD 27 positions (see section 8.5.1) can be displaced hundreds of meters compared to corresponding UTM NAD 83 positions (Welch and Homsey 1997). Such mismatches are, in one sense, better than the much more subtle mismatch caused by confusing two geocentric datums, because the big mismatches are obvious. It is immediately apparent that something is wrong, prompting a search for a solution. Geocentric datum mismatches typically introduce displacements of a meter or less, which might go unnoticed. The following two practices can help. First, data ought to come with metadata that include the reference system, the realization name, the reference epoch, the linear unit, and the coordinate system (XYZ, ENU, LBH, grid), including any projection parameters. Second, GIS projects should include a base layer of authoritative geodetic control markers that can be used as registration points for the other layers. This layer provides a check for all the others so that if one layer comes in displaced, control points can help detect and debug the problem.

Velocity vectors are linear approximations to motions that actually are nonlinear. Therefore, any particular reference frame will eventually become obsolete because the simple linear velocity approximation will cease to be accurate enough. Also, geodesy is evolving rapidly. New methods are being developed that provide better knowledge of the geocenter, of tectonic plate motions, of the Earth's gravity field, of the Earth's orientation with respect to the solar system, and other important geodetic parameters. This better knowledge is often implemented in new reference frames, so the parade of reference frames will not end. Positions of control stations will likely change with every new realization due to these considerations and others. This requires that geomatics professionals be aware of this so that they document the observation epoch of observations and the reference epoch of the datum to which they refer. Using coordinates that refer to different frames is no more sensible than mixing degrees Fahrenheit and degrees Kelvin.

5.7 Your Turn

Problem 5.1. What are the coordinates of the origin of an XYZ coordinate system?

Problem 5.2. What are the XYZ coordinates of a point on the equator and on the prime meridian?

Problem 5.3. What are the XYZ coordinates of a point at the North Pole?

Problem 5.4. What are the XYZ coordinates of a point at the South Pole?

Problem 5.5. What are the XYZ coordinates of a point on the equator and on the international date line?

Problem 5.6. What are the XYZ coordinates of a point on the GRS 80 reference ellipsoid at latitude = 45° N and longitude = 30° W?

Problem 5.7. What are the XYZ coordinates of a point on the International 1924 reference ellipsoid at latitude = 45° N and longitude = 30° W?

Problem 5.8. What is the distance in meters between the XYZ point in problem 5.6 and the XYZ point in problem 5.7?

Problem 5.9. Suppose an airplane was flying due north maintaining a constant 0° E longitude. Which of the three XYZ coordinates would be changing?

Problem 5.10. Suppose an airplane was flying due north maintaining a constant 45° E longitude. Which of the three XYZ coordinates would be changing?

Problem 5.11. Suppose an airplane was flying due north maintaining a constant 90° W longitude. Which of the three XYZ coordinates would be changing?

Problem 5.12. Suppose an airplane was flying due east maintaining a constant 45° N latitude. Which of the three XYZ coordinates would be changing?

Problem 5.13. How far (in meters) is 90° of latitude assuming a spherical Earth with a radius equal to the (i) WGS 84 semimajor axis, (ii) WGS 84 semiminor axis, (iii) average of the WGS 84 semimajor and semiminor axes, and (iv) using a radius equal to $(2a + b)/3$? The length of the WGS 84 meridional quadrant is 10 001 965.730 m.

Problem 5.14. Assuming a spherical Earth with a radius equal to (iv) from problem 5.13, how far (in meters) is $1''$ of latitude?

Problem 5.15. How far (in meters) is 1° of longitude at (i) 0° degrees of latitude, (ii) 45° degrees of latitude, and (iii) 90° degrees of latitude on the WGS 84 reference ellipsoid? (Hint: use reduced latitude.)

5.7. YOUR TURN

Problem 5.16. Inverse between the *ENU* coordinates of the observation station and the forward station in Example 5.3 on page 83 to determine the distance between them. Verify this distance equals the slope distance given in the problem.

Problem 5.17. Inverse between the *XYZ* coordinates of the observation station and the forward station in Example 5.3 on page 83 to determine the distance between them. Verify this distance equals the slope distance given in the problem.

Problem 5.18. Implement the computations in Example 5.3 on page 83 in a spreadsheet.

Problem 5.19. Transform WGS 84 (4.89935°S, 145.96311°E, 2138.9) to WGS 84 *XYZ* coordinates (use the ellipsoidal Eqs. 5.1).

Problem 5.20. Take the answer in Example 5.4 back to AGD.

Problem 5.21. Longitude and latitude alone cannot fully document a position. What needs to be known to fully document a position?

Chapter 6

Distances

Distances are fundamental to mapping and to GIS analyses. For example:

- The forward problem is to determine coordinates of an observed station given the location of the observation station, a distance, and an azimuth.

- The inverse problem is to determine distance and direction given coordinates of two stations.

- One of the more common GIS analyses is to compute distances between points, along multi-lines, and around polygons.

Just as there are many coordinate systems in geomatics, there are many different concepts of distance. In fact, each type of coordinate system brings its own notion of distance, and each type of distance typically means something different than the others.

What is distance? We begin by defining the length, which is called , **arc length** along a possibly bending curve. Suppose we want to know the arc length s of curve C between **A** and **B**, shown in red in Fig. 6.1. As a first try, s can be approximated by a chord length δs_0 between **A** and **B**, with $s^2 \approx \delta s_0^2 = \delta x_0^2 + \delta y_0^2$. A better approximation comes from splitting C between **A** and **B** into two chords, $\delta s_{1,1}$ and $\delta s_{1,2}$. So, $s \approx \delta s_{1,1} + \delta s_{1,2}$. $\delta s_{1,1}$ and $\delta s_{1,2}$ match C more closely than δs_0, so their sum provides a better estimate. Subdividing C into more and more chords makes the chords

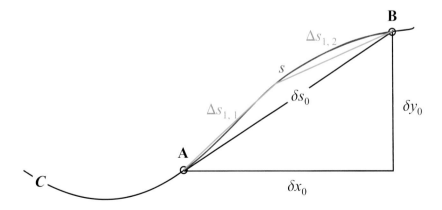

Figure 6.1: Arc length s (in red) is the directed length of a curve between **A** and **B**.

103

smaller, of length s_i, and improves the estimate for s. In the limit for smooth curves, the sum of the chord lengths approaches s as the number of chords approaches infinity: $s = \sum_i^\infty \delta s_i$. When the number of chords has become infinite, the chords become infinitely small and are written as ds. When the summation has become infinite, $\sum$ (sigma) is replaced with an elongated "S", the integral sign:

$$s = \int_{\mathbf{A},\mathbf{B}} ds \qquad (6.1)$$

Arc length depends on direction: $s_{\mathbf{A},\mathbf{B}} = -s_{\mathbf{B},\mathbf{A}}$; distance does not. We define the distance between two points in terms of the magnitude of the arc length along some curve joining them. But, which curve? Obviously, there are an infinite number of curves that can connect any two points. We will define the **distance** between points **A** and **B** to be the length of the shortest curve connecting them. We will denote this curve by the Greek letter γ (gamma) and its length by d. Equation 6.1 often has a simple solution when C is embedded in a surface of constant curvature, such as a plane or a sphere. If **A** and **B** are in a plane, then γ is a straight line and Eq. 6.1 becomes

$$d = [\,(x_{\mathbf{A}} - x_{\mathbf{B}})^2 + (y_{\mathbf{A}} - y_{\mathbf{B}})^2 + (z_{\mathbf{A}} - z_{\mathbf{B}})^2\,]^{1/2} \qquad (6.2)$$

for the three–dimensional case. The nonempty intersection of a plane and a sphere is either a circle or, if they are tangent, a single point. Suppose a plane and a sphere intersect in a circle. If **A** and **B** are on this circle, then C is a circular arc and Eq. 6.1 becomes

$$d = r\,\theta \qquad (6.3)$$

where r is the radius of the circle in the plane, and θ is the angle subtended along the arc in radians. When the plane includes the center of the sphere, the circle is called a **great circle** (section 6.1.4 on page 108). The shortest distance between two points on a sphere is the length of a great circle arc, so γ is the shorter of the two great circle arcs connecting **A** and **B**.

In the above examples **A** and **B** were in either a plane or a sphere. However, real places are not embedded on surfaces; they just exist in space. Suppose **A** and **B** are two points on the Earth's surface. The great circle distance is definitely a meaningful measure of their separation, but they can be conceptualized just as easily as being two points in space without any thought of the Earth. This leads to a key concept:

The type of distance used to describe the separation between points depends on the choice of a surface in which they are thought to be embedded.

In general, the separation between any two points can be described by any type of distance. It's up to the geomatics professional to choose. Choosing the type of distance between two points is an important skill in geomatics because different types mean different things.

If C is embedded in a surface that does not have constant curvature, then γ is a complex curve and Eq. 6.1 does not have a simple form. Distance formulæ for an ellipsoid Earth-model have no closed algebraic form, so devising better ways to compute distances on an ellipsoid remains an active research area.

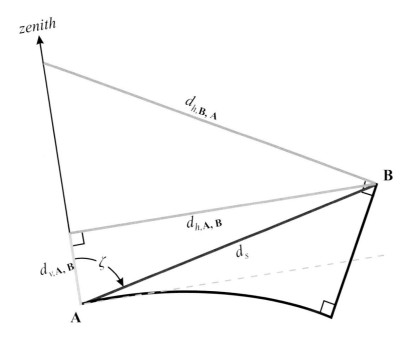

Figure 6.2: The Earth's curvature invalidates the "flat Earth" assumption causing the simplistic relationship between slope distances, zenith angles, and horizontal and vertical distances given in Eq. 3.9 to lead to the horizontal and vertical distance from **A** to **B** in **A**'s *ENU* system not being equal to the horizontal and vertical distance from **B** to **A** in **B**'s *ENU* system.

6.1 Types of Distances

6.1.1 Ground distances

Distances between objects measured along a topographic surface are called **ground distances.** Ground distances can be measured with chains, tapes, wires, or calibrated wheels with odometers. They are needed whenever an analysis depends upon distances on the Earth's surface. This can be the case for highway construction, hydrology, and meteorology, for example. Ground distances can be approximated from digital terrain models (DTM), such as a **triangulated irregular network (TIN)** (Peucker et al. 1978), being a tessellation of planar triangles, called **facets**. A ground distance is computed by determining the intersections of the path on the ground across the TIN's facets. The intersections are line segments, so the ground distance, which is actually an arc length, is approximated by the sum of the lengths of the line segments (Fig. 6.1). The accuracy of such a computed ground distance depends on the density and sampling pattern of the DTM samples, but in all cases, is an underestimation: straight-line segments are always shorter than the curved path along the ground (Blais et al. 1986; Makarovic 1973; Makarovic 1977a; Makarovic 1977b; Makarovic 1984).

6.1.2 Horizontal and vertical distances (*ENU*)

Horizontal and vertical distances were introduced in section 3.2 on page 21. As previously discussed, if the vertical axis of a total station is coincident with the normal of the reference ellipsoid at the observation station, the plane perpendicular to the vertical is the local geodetic horizontal, which is the basis for the *ENU* coordinate system at that point. A horizontal distance is a straight-line distance in the local horizontal plane, and a vertical distance is a straight-line distance perpendicular to the local horizontal plane.

Common sense dictates that distance not depend on direction: $d_{\mathbf{A},\mathbf{B}} = d_{\mathbf{B},\mathbf{A}}$. Within a single *ENU* system, horizontal and vertical distances certainly satisfy this criterion. In fact, considering that an *ENU* system is obtained by translating and rotating an *XYZ* system and neither operation changes the shape of the system, *ENU* distances are identical with their *XYZ* counterparts. That is, if any two points are represented by both *XYZ* and *ENU* coordinates, the slope distance between them is the same regardless of the coordinate system in which it is computed.

But what about comparing horizontal distances between two stations in their own *ENU* systems? The local vertical is unique to the origin, so the local horizontal planes of different *ENU* systems cannot be coincident or even parallel. Therefore, a horizontal distance in one *ENU* system is a slope distance in another, and the triangle inequality (section 3.4.6 on page 27) guarantees that they cannot be the same length. Equation 3.9 on page 35, which is the basis for plane surveying, contained an assumption that the quantities in the equation form a right triangle, and that is not strictly correct. As shown in Fig. 6.2 on the previous page, the horizontal at B is not parallel to that at A. The local horizontal at A (dashed green line in Fig. 6.2 on the preceding page) is not perpendicular to the vertical at B. This causes the computed vertical distance $d_{v,\mathbf{A},\mathbf{B}}$ (in green) to not equal the actual separation of B from the vertical datum (in black). Likewise, $d_{h,\mathbf{A},\mathbf{B}} \neq d_{h,\mathbf{B},\mathbf{A}}$. Furthermore, the geodetic "horizontal" separation between A and B is the black arc, not a horizontal distance at all! Nevertheless, Eq. 3.9 is a good approximation if the separation between the observation station and the forward station is not too great. As discussed in section 4.1 on page 55, although the vertical error grows with the square of the distance, the horizontal error grows only linearly, a much smaller rate.

Horizontal distance discrepancies behave differently for displacements in longitude, latitude, and height. Horizontal distances between points separated in only longitude are not different because the radius of curvature in the prime vertical does not change with longitude. Horizontal distances between points separated in only latitude are different because the meridional-curvature radius does change with latitude, but discrepancy is minute, amounting to 1 mm at a separation of 39' of latitude (71.9 km) starting at the equator of the GRS 80 ellipsoid. Horizontal distances between points separated in ellipsoid height are different for reasons related to the triangle inequality (Fig. 6.2), but the discrepancy becomes concerning much more quickly: for points on the equator of the GRS 80 ellipsoid, 1 m of ellipsoid height difference leads to 1 mm horizontal distance discrepancy at a longitude separation of $21''.5$ (634 m), and 20 m of ellipsoid height difference leads to 1 mm horizontal distance discrepancy at a longitude separation of $10''.5$ (325 m).

Maps compiled with horizontal distances reflect distances on the topographic surface, so such maps are said to be compiled on the ground. Plats (private property maps) are usually compiled with horizontal distances so that the distances shown on the plat between the bounds are the same numbers as the outputs of the distance measuring instrument (this avoids confusion when explaining the plat to clients and lawyers, and is helpful for home improvement projects, too). In principle, apart from strictly longitudinal separations, the two horizontal distances between two bounds are inconsistent, but the inconsistency is usually less than the distance measuring instrument's precision if the topography is not very steep and the separation is not too far.

Horizontal distance discrepancies are seldom problematic for individual-parcel plats, but prob-

6.1. TYPES OF DISTANCES 107

lems can arise when the plats spanning a large area, such as a city, are compiled together into a GIS. GIS is coordinate based, not measurement based, so plats must be represented in a GIS by the coordinates of their features. Each total station setup constitutes its own *ENU* coordinate system, and mixing coordinates from different coordinate systems leads to inconsistencies. One solution to this dilemma was given in chapter 5, where it was shown that *ENU* coordinates can be transformed into geodetic coordinates and, thus, become consistent. In this chapter we take a different approach, one based on reducing observed distances down to a reference ellipsoid. This approach has no coordinate system transformations, so it maintains a sense of plane surveying. Once the distances are reduced to a common reference surface, coordinates can be computed using plane surveying mathematics rather than coordinate system transformations.

6.1.3 Slope distances (XYZ)

As discussed in section 3.2 on page 21, a slope distance is the length of a straight line between two points. Slope distances are computed by inversing in either an *ENU* or an *XYZ* coordinate system (see sections 5.1 and 5.3).

Example 6.1. What is the slope distance between two stations whose NAD 83(CORS96) XYZ coordinates in meters are
ALDRICH = (1 451 508.754, -4 534 331.620, 4 230 245.348)
KERR = (1 451 605.977, -4 534 387.551, 4 230 147.979)

Solution: From Eq. 6.2

$$\begin{aligned} d &= [\,(x_\mathbf{A} - x_\mathbf{K})^2 + (y_\mathbf{A} - y_\mathbf{K})^2 + (z_\mathbf{A} - z_\mathbf{K})^2\,]^{1/2} \\ &= [\,(1\,451\,508.754 - 1\,451\,605.977)^2 + (-4\,534\,331.620 - (-4\,534\,387.551))^2 + \\ &\quad (4\,230\,245.348 - 4\,230\,147.979)^2\,]^{1/2} \\ &= 148.531 \text{ m} \end{aligned}$$

□

The laser beam emitted from an EDM travels in a straight line, ignoring refraction effects in the atmosphere and the extremely small deflections caused by the Earth's gravitational field (Höpcke 1966). However, even if the beam is refracted, its curvature cannot be very large and the length of the refracted beam is very nearly that of the unrefracted beam. Therefore, for most practical purposes, the distance measured by an EDM can be considered to be a slope distance.

Example 6.2. Continuing Example 6.1, KERR's *ENU* coordinates in ALDRICH's system are (75.5427, -127.848, -3.120 93), in meters. Compute the slope and horizontal distances from KERR to ALDRICH in ALDRICH's *ENU* system.
Solution: $d_{s,\mathbf{A},\mathbf{K}} = \sqrt{75.5427^2 + (-127.848)^2 + (-3.120\,93)^2} = 148.531$ m, which is exactly the same as the XYZ slope distance.
The horizontal distance is $d_{h,\mathbf{A},\mathbf{K}} = \sqrt{75.5427^2 + (-127.848)^2} = 148.4981$ m.
□

Example 6.3. ALDRICH's *ENU* coordinates in KERR's system are (-75.5414, 127.848, 3.117 47), in meters. Notice that these are not precisely the negative of KERR's coordinates in ALDRICH's

system. Compute the slope and horizontal distances from KERR to ALDRICH in KERR's system.
Solution: $d_{s,\mathbf{K},\mathbf{A}} = \sqrt{-75.5414^2 + 127.848^2 + 3.11747^2} = 148.531$ m, which is exactly the same as the slope distance in ALDRICH's system and for XYZ.
The horizontal distance is $d_{h,\mathbf{K},\mathbf{A}} = \sqrt{-75.5414^2 + 127.848^2} = 148.4982$ m, a difference of 0.0001 m compared to $d_{h,\mathbf{A},\mathbf{K}}$. The vertical distances differ by 3.117 47 - 3.120 93 = 0.0035 m.
□

6.1.4 Great circle distances (λ, ϕ)

A **great circle** is a circle formed by the intersection of the sphere and a plane such that the plane contains the sphere's center. Great circle distances are popular with navigators, but they are becoming obsolete because modern navigation computations pose no challenges to digital computers even with the ellipsoidally correct formulæ. Important characteristics of great circles can be summarized as follows:

- Two antipodal points on a sphere are connected by infinitely many great circles.

- Two nonantipodal points on a sphere are connected by one and only one great circle.

- A great circle's radius equals that of the sphere itself and thus is maximal. That is to say, no circle on the surface of a sphere has a greater radius than that of a great circle. This implies that circular arc segments of a great circle have minimal curvature, which can be used to prove that these arcs have minimal length.

- The shortest distance between two nonantipodal points on a sphere is the length of the shorter great circle arc joining them.

The length of a great circle arc is given by Eq. 6.3, $d = r\theta$, where r is the radius of the circle (which equals the radius of the sphere), and θ is the angle subtended by the arc in the plane of the great circle. If two points $\mathbf{P}$ and $\mathbf{Q}$ have positions given in XYZ coordinates, those position coordinates can also be interpreted as the coordinates of XYZ vectors from the center of the sphere to the points. Let $\hat{\mathbf{P}}$ and $\hat{\mathbf{Q}}$ denote unit vectors in the direction of $\mathbf{P}$ and $\mathbf{Q}$, respectively. Then, from Eq. A.2 on page 224, $\cos\theta = \hat{\mathbf{P}} \cdot \hat{\mathbf{Q}}$ gives the angle between the vectors in their common plane. Therefore, the great circle distance from $\mathbf{P}$ to $\mathbf{Q}$ is

$$d_{gc} = r \arccos(\hat{\mathbf{P}} \cdot \hat{\mathbf{Q}}) \tag{6.4}$$

with the arccosine function returning an angle in radians.

One common choice for r is 1 (a unit radius). Others come from reference ellipsoids, such as the length of the semimajor axis, an average radius equal to $(2a + b)/3$,[1] a radius such that its product with $\pi/2$ equals the length of a meridional quadrant, and the radius of an authalic sphere. An **authalic sphere** is a sphere whose surface area equals that of some ellipsoid.

Great circle distances can be computed from geographic coordinates. The angle θ can be found using the law of cosines for spherical triangles:

$$\theta = \arccos[\sin\phi_\mathbf{A} \sin\phi_\mathbf{B} + \cos\phi_\mathbf{A} \cos\phi_\mathbf{B} \cos(\lambda_\mathbf{A} - \lambda_\mathbf{B})] \tag{6.5}$$

---

[1] $(2a + b)/3$ is a good approximation of an ellipsoid's average meridional radius of curvature. For the reference ellipsoids given in chapter 4, in the worst case this formula is incorrect (too large) by only 3617.74 m for the Clarke 1880 ellipsoid.

6.1. TYPES OF DISTANCES

where ϕ is latitude and λ is longitude. Substituting Eq. 6.5 into Eq. 6.3 gives

$$d_{gc} = r \cdot \arccos[\sin \phi_\mathbf{A} \sin \phi_\mathbf{B} + \cos \phi_\mathbf{A} \cos \phi_\mathbf{B} \cos(\lambda_\mathbf{A} - \lambda_\mathbf{B})] \tag{6.6}$$

Example 6.4. Assuming a spherical Earth of radius $6\,378\,137$ m, find the great circle distance from $\mathbf{A} = (23°\mathrm{S}, 115°\mathrm{E})$ to $\mathbf{B} = (49°\mathrm{N}, 13°\mathrm{W})$ using Eq. 6.6.
Solution:

$$\begin{aligned} d_{gc} &= r \cdot \arccos[\sin \phi_\mathbf{A} \sin \phi_\mathbf{B} + \cos \phi_\mathbf{A} \cos \phi_\mathbf{B} \cos(\lambda_\mathbf{A} - \lambda_\mathbf{B})] \\ &= 6\,378\,137 \arccos[\sin(-23°)\sin(49°) + \cos(-23°)\cos(49°)\cos(115° - [-13°])] \\ &= 14\,673\,255.86 \text{ m} \end{aligned}$$

The result is deliberately given to more significant digits than would be warranted from whole integer input coordinates, in order to facilitate comparison with the next example.
☐

Equation 6.6 can give incorrect answers for very short lines. This shortcoming is overcome by the haversine formula:

$$d_{gc} = 2r \arcsin\left[\sqrt{\sin^2([\phi_\mathbf{B} - \phi_\mathbf{A}]/2) + \cos \phi_\mathbf{A} \cos \phi_\mathbf{B} \sin^2([\lambda_\mathbf{B} - \lambda_\mathbf{A}]/2)}\right] \tag{6.7}$$

Example 6.5. Assuming a spherical Earth with a radius of $6\,378\,137$ m, find the great circle distance from $\mathbf{P} = (23°\mathrm{S}, 115°\mathrm{E})$ to $\mathbf{B} = (49°\mathrm{N}, 13°\mathrm{W})$ using Eq. 6.7.
Solution:

$$\begin{aligned} d &= 2a \arcsin\left[\sqrt{\sin^2([\phi_\mathbf{B} - \phi_\mathbf{A}]/2) + \cos \phi_\mathbf{A} \cos \phi_\mathbf{B} \sin^2([\lambda_\mathbf{B} - \lambda_\mathbf{A}]/2)}\right] \\ &= 2 \times 6\,378\,137 \arcsin\left[\sqrt{\sin^2([49° - (-23°)]/2) + \cos(-23°)\cos 49° \sin^2([(-13°) - 115°]/2)}\right] \\ &= 14\,673\,255.86 \text{ m} \end{aligned}$$

This is exactly the same answer as the previous example. The distinction between the two formulæ becomes apparent only when the distance is very small, on the order of tens of meters.
☐

6.1.5 Great ellipse distances (λ, ϕ)

Great circle distances are not much used in geodesy, which depends on ellipsoidal Earth models. A great ellipse is the equivalent of the great circle for an ellipsoid (Bowring 1996). A **great ellipse** is the ellipse formed by the intersection of an ellipsoid and a plane containing the ellipsoid's center. Unlike a great circle, a great ellipse generally does not provide the shortest curve between two points, the exceptions being the equator or any meridian. Recall that the line segment from an ellipse's center to the ellipse is generally not perpendicular to the ellipse (Fig. 4.3 on page 58). A great circle's constant maximal radius is the property that establishes its shorter arc as the shortest length curve, because maximal radius implies least curvature, which makes it the most nearly straight curve on a sphere. A curve's radius of curvature at a point is maximal when the vector normal to the curve at that point is parallel to the vector normal to the surface in which that

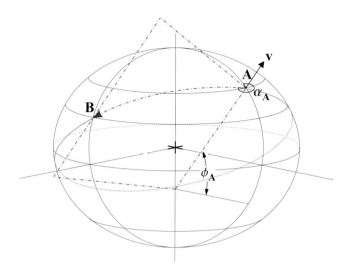

Figure 6.3: The normal section at **A** oriented according to $\alpha_\mathbf{A}$. The normal section arc from **A** to **B** is dashed red. The entire normal section is the arc plus the rest of the ellipse shown in green. The sectioning plane contains the normal vector **v** at **A**, is oriented according to $\alpha_\mathbf{V}$, and does not contain the ellipsoid's origin.

curve is embedded (Pressley 2007). This is always true for a great circle, but since a great ellipse's normal vector is not always normal to the ellipsoid, it is not true in general for great ellipses.

Great ellipse arcs have been put forward as a suitable candidate to represent "straight lines" in a GIS, such as polygon edges and multiline segments (Kallay 2007). The issue arises because straight lines in a GIS represent curves on the Earth, which is modeled with an ellipsoid, but there are no straight lines on an ellipsoid. The geodesic (see section 6.1.8 on page 113) is the most rigorous choice, but it is difficult to compute. Normal section arcs (see below) are not unique: at least two normal section arcs connect any two points, which would disqualify them from a practical perspective. Great ellipse arcs are unique and are, in fact, quite close to being the shortest curves.

6.1.6 Normal section distances (λ, ϕ)

Normal sections (introduced in section 4.4.3 on page 69) are ellipses, but generally not great ellipses. Consider points **A** and **B** on a reference ellipsoid, not both on a meridian or the equator. The great ellipse from **A** to **B** would not be the normal section from **A** to **B** because the surface normal vector at **A** will not be in the plane containing the ellipsoid's origin at **A** (Fig. 6.3). Great ellipses are normal sections only for the equator and the meridians. The equator is the only ellipsoidal normal section that is a parallel of latitude.

Figure 6.4 on the facing page shows the normal section arc at **B** containing **A**, which is a different curve than the normal section arc at **A** containing **B**. The geodesic lies between the normal section arcs (Fig. 6.5). Two different normal section arcs will connect two points if they are not on the same parallel nor on the same meridian because the surface normal vectors at the points are not coplanar; they are skew to each other. If the two normals are skew, then they cannot both be in the same sectioning plane. This **skew of the normals** is caused by an eccentricity greater than zero; the greater the eccentricity, the worse the skewness.

6.1. TYPES OF DISTANCES

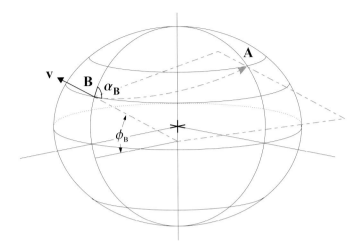

Figure 6.4: The normal section at **B** oriented according to $\alpha_\mathbf{B}$. The sectioning plane contains the normal vector **v** at **B**, is oriented according to $\alpha_\mathbf{B}$, and does not contain the ellipsoid's origin.

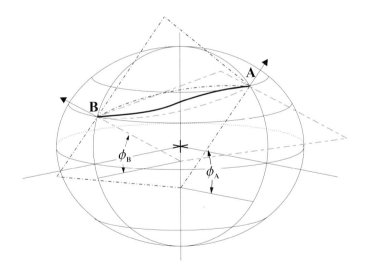

Figure 6.5: A composite picture putting Fig. 6.3 with Fig. 6.4. The normal vectors intersect the z-axis in different locations, so they are not coplanar. The geodesic from **A** to **B** is shown between the normal sections as a solid black curve.

When an observed station **B** is not on the reference ellipsoid, it is not contained in the sectioning plane at the observing station **A**. This causes a small angular discrepancy between the observed azimuth α' and the geodesic azimuth α (Rapp 1989a, p. 61–63 and Bomford 1962, p. 94). The relationship is

$$\phi_m = (\phi_\mathbf{A} - \phi_\mathbf{B})/2$$
$$M_m = M(\phi_m)$$
$$\alpha = \alpha' + \frac{h_\mathbf{B}}{2M_m}\epsilon^2 \cos^2 \phi_m \sin 2\alpha' \tag{6.8}$$

where ϵ is first eccentricity, ϕ_m is the average geodetic latitude of the stations, M_m is the meridional radius at ϕ_m, and $h_\mathbf{B}$ is the ellipsoid height of **B**. This correction is always small.

Example 6.6. Station **A** at $\phi_\mathbf{A} = 45°$ observes station **B** at $\phi_\mathbf{B} = 45.01°$ and $h_\mathbf{B} = 3000$ m at an observed azimuth of $\alpha' = 45°$. Using the GRS 80 reference ellipsoid, what is the azimuth after applying the correction for the skew of the normals?
Solution: $\phi_m = 45.005°$. From Eq. 4.7

$$M_m = M(\phi_m) = \frac{a(1-\epsilon^2)}{(1-e^2 \sin^2 \phi_m)^{3/2}}$$
$$= 6\,367\,387.414 \text{ m}$$

Then

$$\alpha = \alpha' + \frac{h_\mathbf{B}}{2M_m}\epsilon^2 \cos^2 \phi_m \sin 2\alpha'$$
$$= 45° + \frac{3000}{2 \times 6\,367\,387.414} 0.006\,694\,380\,023 \cos^2 45.005° \sin 90°$$
$$= 45°\, 0'\, 0\rlap{.}''16$$

□

For geodesy on the Earth, Bomford (1962, p. 102) notes, "The length of the normal section differs from that of the geodesic by under 1 in 150 000 000 in a side of 3000 km, and the difference can always be ignored." Thus, normal section distance is practically equivalent to geodesic distance. Normal section formulæ can be found in Bowring (1971).

6.1.7 Curve of alignment (λ, ϕ)

Thomas (1979, p. 66) defines a **curve of alignment** to be "...the locus of a point on the spheroid which moves so that the plane through it and two fixed points on the spheroid is normal to the surface at the moving point." Such a locus is called a curve of alignment because it would be the trace of an instrument moved between two fixed points so as to keep the telescope aimed at both stations throughout the path between the points. A curve of alignment can be visualized as follows: envision a straight pipe joining the two fixed points (the pipe is inside the ellipsoid); attach a collar around the pipe so that the collar can slide along the pipe's length; attach an unbendable straight wire to the collar perpendicularly to the pipe's axis; place a freely sliding plate on the reference

6.1. TYPES OF DISTANCES 113

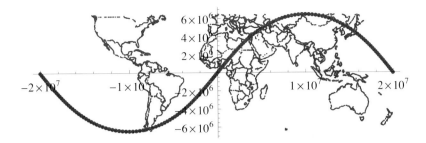

Figure 6.6: Geodesics achieve an absolute maximum latitude, and they oscillate between that latitude in the Northern and Southern hemispheres.

ellipsoid; drill a hole perpendicularly through the plate, and pass the wire through the hole (this constrains the wire to remain normal to the ellipsoid). The trace of the intersection of the wire and the ellipsoid between the points is the curve of alignment. The curve of alignment is not a geodesic, although it is very close. The argument proving it is not a geodesic is based on a concept from differential geometry (geodesic curvature) that is beyond the scope of this book.

6.1.8 Geodesic curves (λ, ϕ)

According to Pressley (2007), "Geodesics are the curves in a surface that a bug living in the surface would perceive to be straight." Pressley's bug has a very limited field of view; it has no notion that it lives on a curved surface because the surface's radius of curvature is so large compared to the size of the bug. The bug does, however, have a clear notion of turning to one side or the other. So, for the moment, let's say that if the bug walks without turning, it is following a geodesic. A straight line is a geodesic for a plane, and a great circle arc is a geodesic for a sphere. A normal section might seem to be a geodesic because it seems to somehow be a straight-line path. After all, the normal section lies in the sectioning plane, so following that curve would seem to not involve turning from side to side. However, a normal sectioning plane is normal to an ellipsoid only at one point. If the bug were quite sensitive to whether it was leaning from side to side, which is one way how organisms perceive themselves turning, the bug would feel as though it was not walking a straight line because following the normal section would cause it to follow a curve whose normal vector is not everywhere perpendicular to the surface being walked on. This is why normal sections are not geodesics. Geodesics are curves whose instantaneous sectioning plane remains normal to the surface throughout their length.

Important features of geodesic curves can be summarized as follows (Thomas 1970):

- Geodesics are specified by a fixed point on the reference ellipsoid and an azimuth.

- Meridians are geodesic curves; all geodesics that include the poles are meridians. The equator is the only parallel that is a geodesic.

- Geodesics can be defined as the solution of a system of three differential equations (Thomas

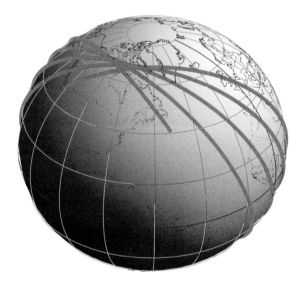

Figure 6.7: Geodesic segments on an ellipsoid do not end where they began after making a complete transit of the reference ellipsoid. The geodesic segment starting in Seattle, after completing a circuit, passes west of where it began.

6.1. TYPES OF DISTANCES

1970, p. 28):

$$ds \cos \alpha = M d\phi \tag{6.9a}$$
$$ds \sin \alpha = N \cos \phi \, d\lambda \tag{6.9b}$$
$$d\alpha = \sin \phi \, d\lambda \tag{6.9c}$$

where M and N are radii of curvature as usual, α is the geodesic's azimuth at the specified fixed point, and s is arc length along the geodesic. Arc length can be positive or negative, so geodesics can be followed forwards (positive arc length) or backwards (negative arc length). Some authors call a specific value of s a "mile marker" (Rollins 2010). Equation 6.9a expresses the change in distance along a meridian ($M d\phi$) caused by moving an infinitesimal distance ds along a curve (not necessarily a geodesic) oriented at azimuth α. Equation 6.9b expresses the change in distance along a parallel ($N \cos \phi \, d\lambda$) caused by moving an infinitesimal distance ds along a curve (not necessarily a geodesic) oriented at azimuth α. Equation 6.9c constrains the curve to be a geodesic. Choosing other equations instead of Eq. 6.9c yields different curves; for example, setting dα to a constant yields a loxodrome. A **loxodrome** (or rhumb line) is a curve that intersects all meridians at the same angle.

- Geodesics achieve an absolute maximum latitude, and they oscillate between that latitude in the Northern and Southern hemispheres (Fig. 6.6 on page 113).

- There is a number called Clairaut's constant c that has the same value for all points on the same geodesic:

$$c = \cos \beta \sin \alpha = \cos \beta_{\max} = \frac{\sin \alpha \cos \phi}{\sqrt{1 - \epsilon^2 \sin^2 \phi}} \tag{6.10}$$

where $\beta_{\max}$ is the maximum reduced latitude achieved by the geodesic. For the direct problem, c can be computed at the given starting point (λ_1, ϕ_1) with initial azimuth α_1 by the last term of Eq. 6.10 (Sjöberg 2007; Rollins 2010).

- A consequence of Clairaut's relationship is that all geodesics that pass through a particular geodetic latitude at a particular azimuth have identical shape regardless of their longitude at that latitude.

- Geodesic-curve segments that complete a full circuit around an ellipsoid with eccentricity greater than zero do not return to where they started (Fig. 6.7 on the facing page). This is caused by the eccentricity of the ellipsoid; a geodesic on a sphere (a great circle) does return to the point from which it began. The figure shows a single geodesic-curve segment that has wrapped multiple times around an ellipsoid with a rather extreme eccentricity of $\epsilon = 0.2$, which greatly exaggerates how far the segment's starting point is from where it completes a circuit. Unlike great ellipses, normal sections, and curves of alignment, ellipsoidal geodesics are generally curves of infinite length.

- An infinite number of geodesics connect the poles (the meridians). Otherwise, there are two geodesic segments connecting two fixed points on an ellipsoid; call the shorter one the forward segment. The shortest curve connecting two fixed points on a reference ellipsoid is a forward geodesic segment.

- A segment of a geodesic has different azimuths at its starting and ending points. The former is called the **forward azimuth** and the latter is called the **back azimuth**. Back azimuths are sometimes defined to be the direction opposite that defined here.

- Two nonantipodal points can lie on two forward geodesics if the points are both on the equator and are nearly antipodal. To envision this, imagine the geodesic for two points fairly close to each other on the equator; the equator is their geodesic. The arc length of a meridian from the equator to a pole is called **the length of a meridional quadrant**. The length of a meridional quadrant is less than a quarter of the equator's length because reference ellipsoids are oblate. Therefore, antipodal points on the equator are on the meridian geodesic connecting them, but they remain on the equatorial geodesic, as well. At a particular separation in longitude equal to $\pi\sqrt{1-\epsilon^2}$ radians, called the **liftoff point**, the equator ceases to be the shortest geodesic between two equatorial points. At this point, the shortest geodesic "lifts off" the equator until it coincides with a meridian when the points are antipodal. In Fig. 6.8 the equator is shown in red and the geodesic between Seattle and London in orange. The geodesic crosses the equator at two points, and the orange geodesic is the "liftoff" geodesic between them (Thien 1967; Thomas 1970).

Geodesic curves are the primary object of interest for the geodesist who is solving either the direct or inverse problem, or for any geomatician whose needs to solve these problems for points separated by more than the limit for normal sections. The additional computational complexity of geodesic curves places their implementation mostly into the comfort zone of professional programmers and geodesists.

A great deal of attention has been given to finding the properties of geodesics on ellipsoids. McCleary (1994), Casey (1996), and Pressley (2007) provide theoretical treatments. A comprehensive treatment for geodesy was given in Thomas (1970), with another detailed discussion presented in Rapp (1989a). Additionally, Hooijberg (1997) provides a detailed development of the problem of finding geodesic distance and addresses the implementation details discussed by Kivioja (1971). There are many papers, including Rainsford (1949a), Rainsford (1949b), Rainsford (1955), ACIC (1959), Sodano (1965), Bowring (1969), Bowring (1972), Vincenty (1971), Vincenty (1975), Jank and Kivioja (1980), Murphy (1981), Bowring (1985), Day (1987), Danielsen (1994), Vassallo and Secci (1995), and Rollins (2010).

Direct and inverse problems of geodesics

Clarke (1880) first presented the original formulæ followed by several methods that were compared in ACIC (1959). Robbins (1962) developed these further, giving formulæ with a maximum error of 1/180 m over 1600 km. Bowring (1971) extended Robbins (1962) so that distances up to 6000 miles could be computed. Bomford (1962) provides a review of several approaches (p. 108–110).

The inverse problem of geodesics is as follows: given point $\mathbf{A} = (\lambda_\mathbf{A}, \phi_\mathbf{A})$ and point $\mathbf{B} = (\lambda_\mathbf{B}, \phi_\mathbf{B})$ on a reference ellipsoid, determine (i) the arc length of the forward geodesic segment s_G from $\mathbf{A}$ to $\mathbf{B}$, (ii) the forward azimuth $\alpha_\mathbf{A}$, and (iii) the backward azimuth $\alpha_\mathbf{B}$. Notation: a and b denote lengths of semimajor and semiminor axes, N denotes radius of curvature in the prime vertical, ϵ denotes first eccentricity, and $\delta\lambda = \lambda_\mathbf{B} - \lambda_\mathbf{A}$. There are the following cases:

1. Both points are on the equator and are closer than the liftoff point (i.e., $|\delta\lambda| \leq \pi\sqrt{1-\epsilon^2}$):
   (i) $s_G = a\,\delta\lambda$ (because the equator is a circle)
   (ii, iii) $\alpha_\mathbf{A} = \alpha_\mathbf{B} = \pm\pi/2$, taking their sign from $\delta\lambda$

6.1. TYPES OF DISTANCES

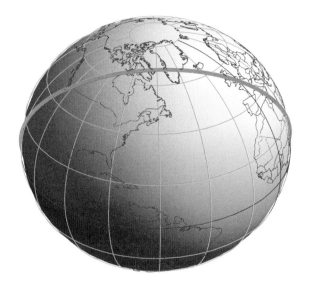

Figure 6.8: The geodesic from Seattle to London passes north of 60° N.

2. Both points are on the equator and farther apart than the liftoff point, but not antipodal (i.e., $\pi\sqrt{1-\epsilon^2} < |\delta\lambda| < \pi/2$): use the methods in Thien (1967) or Rollins (2010).

3. Meridional case (i.e., $\lambda_\mathbf{A} = \lambda_\mathbf{B}$):
   (i) s_G is the length of a segment of a meridian, which has the form

$$s_G = a(1-\epsilon^2) \int_{\phi_\mathbf{A}}^{\phi_\mathbf{B}} (1 - \epsilon^2 \sin^2 \phi)^{-3/2} d\phi \qquad (6.11)$$

This integral is a special case of the Elliptic Integral of the 3rd kind, which is known to not have a closed-form algebraic solution. Many modern computational environments, such as *Mathematica* (Wolfram 1999), support elliptic integrals as built-in functions, so this equation can be evaluated as a closed-form algebraic expression (i.e., not depending on numeric integration methods) in such environments. Geodesics that transit the poles create additional complications.
   (ii, iii) $\alpha_\mathbf{A} = \alpha_\mathbf{B} = 0$ if $\phi_\mathbf{B} > \phi_\mathbf{A}$, and $\alpha_\mathbf{A} = \alpha_\mathbf{B} = \pi$ if $\phi_\mathbf{B} < \phi_\mathbf{A}$

4. General case:
   (i) Following Thomas (1979), Rollins (2010) derived general integrals for change of longitude

and geodesic arc length on a reference ellipsoid:

$$\lambda_B - \lambda_A = \frac{c(1-\epsilon^2)}{\sqrt{1-c^2\epsilon^2}} \int_{\theta_A}^{\theta_B} \frac{1}{(1-k^2\sin^2\theta)\sqrt{1-k^2\epsilon^2\sin^2\theta}} d\theta \qquad (6.12)$$

$$s_B - s_A = \frac{a(1-\epsilon^2)}{\sqrt{1-c^2\epsilon^2}} \int_{\theta_A}^{\theta_B} (1-k^2\epsilon^2\sin^2\theta)^{-3/2} d\theta \qquad (6.13)$$

where s_A and s_B are arc length mile markers along the geodesic segment at **A** and **B**, c is Clairaut's constant, $\theta = \arcsin([\sin\phi]/k)$ is a change of variable replacing latitude, and k is a constant defined as

$$k = \sqrt{\frac{1-c^2}{1-c^2\epsilon^2}}$$

The integral in Eq. 6.13 is also a special case of the Elliptic Integral of the 3rd kind. For $\alpha_1 = 0$, $c = 0$ and $k = 1$, so Eq. 6.13 simplifies to Eq. 6.11.

(ii, iii) Rollins (2010) provides methods to compute α_A and α_B.

The direct problem of geodesics is as follows: given point $\mathbf{A} = (\lambda_A, \phi_A)$, a forward azimuth α, and a length of geodesic arc s_G, determine point $\mathbf{B} = (\lambda_B, \phi_B)$ on the forward geodesic. The direct problem has cases similar to the inverse problem, excepting the liftoff problem: there is always only one geodesic implied by the given parameters.

The direct and inverse problems of geodesics can be implemented without using elliptic integrals by using algebraic approximations. Bowring (1996) provides a method for which he claims, "The inverse problem for all possible geodesics on the ellipsoid is solved in ways that are selected by the program in a manner appropriate to any two given end positions. The comparatively simple total inverse solution for the great elliptic is also given." In fact, Bowring's method is not as general as Rollins's because (i) Rollins allows longitude on the entire real line, whereas Bowring confines $\lambda_B - \lambda_A$ to be between 0 and 2π, and (ii) Rollins's equations handle all the cases above, excepting the simple equatorial case, but including transiting the poles. Bowring's formulæ are extensive, requiring two pages of notation, so they will not be recapitulated here. In contrast, Rollins's method requires only computing c and k once for a particular geodesic; for the inverse problem, evaluating Eq. 6.13 with two invocations of an elliptic integral function; for the direct problem, evaluating Eq. 6.12 with two invocations of an elliptic integral function, and using a numeric root-finding technique to iterate over Eq. 6.13 for θ_B. Rollins's method is by far the simpler approach. The geodesics in Fig. 6.6, 6.7, and 6.8 were drawn using Rollins's method.

6.2 Distance Reductions

In chapter 5, coordinate system transformations were applied to total station observations to transform *ENU* coordinates to geodetic coordinates of the forward station, which can be projected to grid coordinates with map projections (chapter 8). There is another way. This section presents a series of reductions that reduce observed quantities to normal section distances, and then to grid distances and angles. The reduced quantities can be used with plane surveying mathematics to compute grid coordinates, which is our primary goal.

6.2. DISTANCE REDUCTIONS

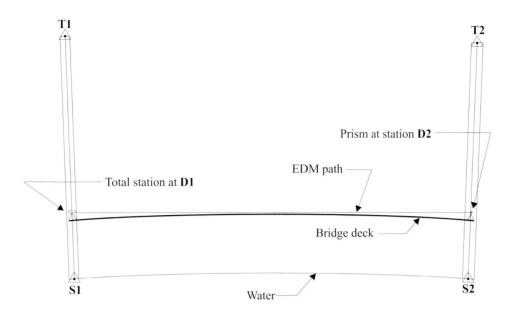

Figure 6.9: Simplistic rendering of the Verrazano Narrows Bridge showing the two towers, the bridge deck, and the water. This illustrates straight-line EDM distances (chords) versus arcs on the ellipsoid.

6.2.1 Chord length versus arc length

In the past, there have been long baselines measured with chains that provided the linear basis for large-extent geodetic surveys. In the original British survey of India, the initial baselines were mechanically flattened to remove topographic variation, and the chains were placed on leveled tables to remove the catenary shape from the chain (Keay 2000). Some of the baselines were nearly ten miles long and took many months to measure. Today, one could simply set up a sufficiently powerful EDM and prism at the endpoints of the baseline and make what might seem to be the same measurement in a matter of seconds, assuming intervisibility of the stations. Conceptually, however, these are really not measurements of the same thing at all. Let us consider the difference between straight-line distances and distances measured along the surface of a reference ellipsoid (i.e., on an arc) by looking at the Verrazano Narrows Bridge, located at the mouth of upper New York Bay (Fig. 6.9). The deck is 4260 ft long and assumed to be at a constant 228 ft (69.5 m) above mean high water, and the towers are 693 ft tall. The deck of the bridge and the surface of the bay are shown as arcs to emphasize the curvature of the Earth. If the bridge's deck is level, then it has to follow the Earth's curvature. Thus, level lines on the Earth's surface are not straight. Suppose a total station and a prism reflector are set up as shown in Fig. 6.9. The straight line between the total station and the prism depicts the ideal path that the beam from the EDM would take. We will now consider the difference between the length of the deck and the distance measured by the EDM.

Consider a circle whose radius equals the semimajor axis of the GRS 80 ellipsoid (6 378 137 m) plus 69.5 m for the height of the deck. Denote this distance as $r = 6\,378\,206.5$ m (Fig. 6.10 on the following page). Let **O** denote the point at the center of the circle and let **A** and **B** denote two points on the circle. From Eq. 6.3, the arc length between **A** and **B** is $d_{arc} = r\,\theta$. The length of the

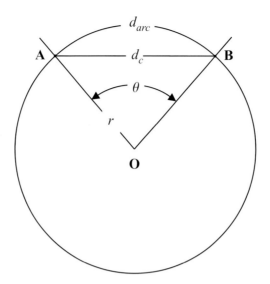

Figure 6.10: Chord distances versus arc distances.

chord $\overline{\mathbf{AB}} = d_c$ is given by the law of cosines (Eq. 3.2) as $d_c = \sqrt{2r^2 - 2r^2 \cos\theta}$. For the Verrazano Narrows Bridge, d_{arc} was given as 4260 ft, so

$$\begin{aligned} \theta &= d_{arc}/r \\ &= (4260\,\text{ft} \times 0.3048\,\text{m/ft})/6\,378\,206.5\,\text{m} \\ &= 0.000\,203\,576 \text{ radians} \end{aligned}$$ (6.14)

Then

$$\begin{aligned} d_c &= \sqrt{2r^2 - 2r^2 \cos\theta} \\ &= \sqrt{2 \times 6\,378\,206.5^2 - 2 \times 6\,378\,206.5^2 \cos(0.000\,203\,576)} \\ &= 1298.450\,595\,\text{m}/(0.3048\,\text{m/ft}) \\ &= 4259.999\,99\,\text{ft} \end{aligned}$$

which is a minute difference. Table 6.1 on the next page lists increasing values of d_{arc} and the corresponding value of d_c at sea level for comparison. As would be expected, the chord distance is marginally shorter than the arc distance. At 10 km the difference is roughly 1 mm, and at 100 km the difference is roughly 1 m.

6.2.2 Reduced distances on the sphere

Two pairs of stations separated only in height are not separated by the same horizontal distance (Clark 1963; Heiskanen and Moritz 1967; Bomford 1962). Suppose there are two stations **S1** and **S2** at sea level as shown in Fig. 6.9. Then imagine that there are two other stations **D1** and **D2** exactly above **S1** and **S2** at the height of the bridge deck. That is, **S1** has the same latitude and longitude as **D1**, and **S2** has the same latitude and longitude as **D2**, but **D1** and **D2** are at a height of 228 ft rather than at sea level. The arc distance between **S1** and **S2** is less than the arc

6.2. DISTANCE REDUCTIONS

distance between **D1** and **D2**. Both pairs of stations are on arcs, but the radius for **D1** and **D2** is greater than that for **S1** and **S2**.

The arc distance between **D1** and **D2** was given as 4260 ft, or 1 298.448 m. To find the arc distance between **S1** and **S2**, apply Eq. 6.3 with a radius that does not include the height of the deck. θ is supplied from Eq. 6.14, so

$$\begin{aligned}
d_{\mathbf{S1},\mathbf{S2}} &= a\,\theta \\
&= 6\,378\,137\,\mathrm{m} \times 0.000\,203\,576 \\
&= 1298.44\,\mathrm{m} \times 3.280\,84\,\mathrm{ft/m} \\
&= 4259.96\,\mathrm{ft}
\end{aligned}$$

The distance along the deck is 0.04 ft (0.49 in) longer than the distance across the surface of the bay. The towers stand 693 ft high, so their tops are separated by 4260.12 ft, which is greater than the distance on the water by 1.45 in. These differences are enough to be of concern, and they are taken into account in the design of such large structures.

For mapping, the situation is usually the opposite. Consider a land survey done in Bogota, Colombia, at an elevation of 2600 m. Suppose there are two stations at this elevation connected by a baseline of 10 km measured on the ground. This same baseline measured on an Earth-sized sphere would measure only 9995.94 m. Table 6.2 on the following page shows separations of a 10 km baseline's endpoints, starting on the surface of a sphere whose radius equals the length of the GRS 80 semimajor axis, and increasing in elevation up to 5000 m. Unlike the distances given in Table 6.1, these distances increase significantly with height. Distances on a reference ellipsoid are said to be the **reduced** equivalent of the ground distances they correspond with. Reduced distances are often smaller than ground distances, but not necessarily: if one measured distances between points below the reference ellipsoid (essentially below mean sea level), then the reduction would actually increase the distance.

Unlike the discrepancy between arc length and chord length, reduction to a reference ellipsoid cannot usually be ignored for at least two reasons. First, distances reported by GNSS postprocessing software are usually given on the surface of the reference ellipsoid, not on the ground. Second, maps compiled in an absolute coordinate system, such as the State Plane Coordinate System, are typically compiled from reduced distances. Significant inconsistencies can arise when a surveyor uses unreduced distances to compile a map in, say, the State Plane Coordinate System or compares ground distances with GNSS-derived distances (see below).

Table 6.1: Arc distances d_{arc} and equivalent chord distances d_c on a sphere with a radius equal to the semimajor axis of GRS 80. Distances are given in meters and d_c is rounded to the nearest millimeter.

d_{arc}	1500	10 000	100 000	1 000 000
d_c	1500.000	9 999.999	99 998.976	998 976.098
$d_{arc} - d_c$	< 0.001	0.001	1.024	1023.901
d_{arc}/d_c	1.000 000 003	1.000 000 102	1.000 010 242	1.001 024 951

Table 6.2: Lengths of baselines separating stations at identical latitude and longitude, but at different heights.

Height (m)	Baseline (m)
0	10 000.000
100	10 000.157
500	10 000.784
1000	10 001.568
1500	10 002.352
2000	10 003.136
3000	10 004.704
4000	10 006.271
5000	10 007.839

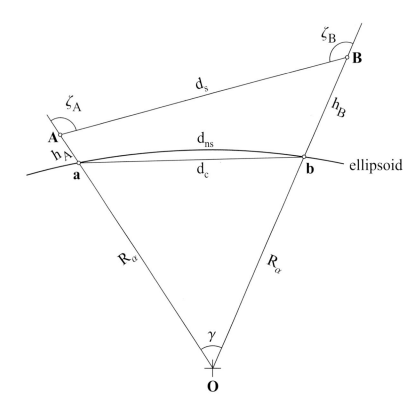

Figure 6.11: Reducing the slope distance d_s to a normal section chord d_c and then to the normal section d_{ns} (in reality, ellipsoid heights are extremely small compared to the radius of curvature, and radii of curvature are normal to the reference ellipsoid).

6.2.3 Normal section chords

Reducing slope distances to the lengths of normal section arcs proceeds as follows: stations **A** and **B** have ellipsoid heights $h_\mathbf{A}$ and $h_\mathbf{B}$, respectively (Fig. 6.11). $\overline{R}_\alpha = (R_{\alpha_{\mathbf{A},\mathbf{B}}} + R_{\alpha_{\mathbf{B},\mathbf{A}}})/2$ is the average radius of curvature in the normal section (Eq. 4.9). To reduce slope distance d_s to the length of the normal section arc, we first need to reduce it to the length of the normal section chord d_c (Heiskanen and Moritz 1967, p. 192):

$$d_c = \sqrt{\frac{d_s^2 - \delta h^2}{(1 + \frac{h_\mathbf{A}}{\overline{R}_\alpha})(1 + \frac{h_\mathbf{B}}{\overline{R}_\alpha})}} \tag{6.15}$$

where $\delta h^2 = (h_\mathbf{B} - h_\mathbf{A})^2$. The length of the normal section arc d_{ns} is

$$d_{ns} = 2\overline{R}_\alpha \arcsin(\frac{d_c}{2\overline{R}_\alpha}) \tag{6.16}$$

Equations 6.15 and 6.16 involve the ellipsoid heights of the two stations. However, these equations are often used in the forward problem, so determining the coordinates of **B** is the task at hand and $h_\mathbf{B}$ is unknown. This problem can be addressed with trigonometric leveling.

6.2.4 Trigonometric leveling with ellipsoid heights

Section 3.6.2 on page 37 introduced the idea of determining height differences with a total station, a process known as trigonometric heighting (or trigonometric leveling). If vertical distances are computed assuming the Earth is flat, the mathematics are very simple: the change in height is the slope distance multiplied by the cosine of the zenith angle. The situation is more complex on the ellipsoid. Suppose we are given the geodetic coordinates of station **A**, including ellipsoid height, reciprocally observed zenith angles $\zeta_\mathbf{A}$ and $\zeta_\mathbf{B}$ reduced to the mark per Eq. 3.12, and the slope distance d_s to an observed station **B** (Fig. 6.11 on the facing page). We can use the following equation to find **B**'s ellipsoid height $h_\mathbf{B}$ (Heiskanen and Moritz 1967, p. 175):

$$h_\mathbf{B} - h_\mathbf{A} = d_{ns}(1 + \frac{\overline{h}}{\overline{R}_\alpha} + \frac{d_s^2}{12\overline{R}_\alpha^2})\tan\frac{\zeta_\mathbf{B} - \zeta_\mathbf{A}}{2} \tag{6.17}$$

where $\overline{h} = \frac{1}{2}(h_\mathbf{A} + h_\mathbf{B})$, d_{ns} is d_s reduced to a normal section per Eq. 6.15 and 6.16, and $\overline{R}_\alpha$ is the mean radius of curvature in the normal section between the stations.

Unless the coordinates of the observation station were determined with GNSS positioning, its height far more likely will be an elevation, which is referred to the geoid (essentially mean sea level) rather than a reference ellipsoid. Orthometric heights are related (to a very close approximation) to ellipsoid heights by

$$h = N + H$$

where H is orthometric height, N is geoid height (see sections 9.3 and 10.1.3; not to be confused with the radius of curvature in the prime vertical), and h is ellipsoid height. N can be on the order of 100 m and is often on the order of dozens of meters.

Equation 6.17 involves $h_\mathbf{B}$ on both sides of the equal sign in such a way that it cannot be algebraically isolated. Therefore, Eq. 6.17 requires iteration. First, an initial, reasonable value, such as $d_s \cos\zeta_\mathbf{A}$, is assigned to $h_\mathbf{B}$. With this estimate, a new value for $h_\mathbf{B}$ is computed, which

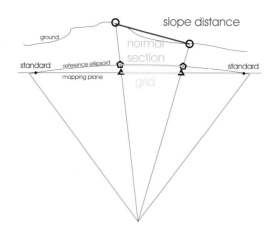

Figure 6.12: Slope distance reduced to the length of a normal section arc and then to a map projection plane. Planimetric distance is less than the normal section distance because the line falls between the standard lines.

then becomes the next value on the right-hand side, and so on until the value for $h_\mathbf{B}$ converges to a solution.

Equation 6.17 involves d_{ns}, which is unknown if we are solving the forward problem. Therefore, solving these equations with full rigor requires a two-step iteration: first iterate Eq. 6.17 to find an initial $h_\mathbf{B}$ with an initial guess for d_{ns}, and then use the initial $h_\mathbf{B}$ to find the next d_{ns} using Eq. 6.15 and 6.16, and so on. This illustrates some of the difficulties of solving the forward problem with full rigor if coordinate system transformations or the ellipsoid forward equations for the geodesic are not used.

Making the simplifying assumption of an appropriately sized, spherical Earth eliminates the iterations. Stem (1995) offers a simplified formula:

$$d_{ns} = d_{\overline{H}} \left(\frac{R}{R + \overline{N} + \overline{H}} \right) \tag{6.18}$$

where $\overline{H}$ is the average orthometric height of the line, $d_{\overline{H}}$ is the horizontal distance between the stations at the average orthometric height of the line, $\overline{N}$ is the average geoid height for the line, and $R = 6\,372\,000$ m for NAD 83, being the mean radius of GRS 80 to the nearest kilometer. One value for $\overline{H}$ and one for $\overline{N}$ can be chosen for an entire job if the site is not too large or does not have too much topographic relief. The quantity in parentheses is sometimes called an **elevation factor**.

6.2.5 Planimetric (grid) distances

Planimetric distances are computed with the two–dimensional version of Eq. 6.2. For two points **A** and **B** in a map projection grid

$$d_p = [\delta x^2 + \delta y^2]^{1/2} \tag{6.19}$$

where $\delta x = x_\mathbf{A} - x_\mathbf{B}$ and $\delta y = y_\mathbf{A} - y_\mathbf{B}$. The arc length of a curve on a reference ellipsoid will generally not equal the arc length of the projected image of that curve. The difference between the ellipsoidal and projected arc lengths is associated with the **projection scale factor** k, which is

6.2. DISTANCE REDUCTIONS

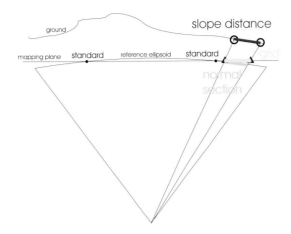

Figure 6.13: Slope distance reduced to the length of a normal section arc and then to a map projection plane. Planimetric distance is greater than the normal section distance because the line falls outside the standards.

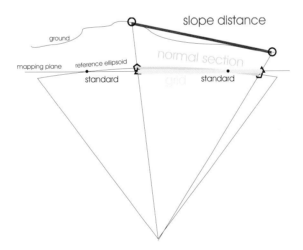

Figure 6.14: Slope distance reduced to the length of a normal section arc and then to a map projection plane. Planimetric distance in part expanded and in part contracted because the line falls on both sides of the standards.

the ratio of an infinitely short planimetric arc length $\mathrm{d}s_p$ to an infinitely short arc length on the ellipsoid $\mathrm{d}s_e$: $k = \mathrm{d}s_p/\mathrm{d}s_e$ and so $\mathrm{d}s_e = \mathrm{d}s_p/k$. Suppose a curve $\mathcal{C}_e$ on the ellipsoid has an image $\mathcal{C}_p$ under some map projection. Let d_e denote the length of $\mathcal{C}_e$, and d_p denote the length of $\mathcal{C}_p$. From Eq. 6.1

$$d_e = \int_{\mathcal{C}_e} \mathrm{d}s_e$$

Switching to integrating over $\mathcal{C}_p$ and substituting $\mathrm{d}s_e = \mathrm{d}s_p/k$ gives

$$d_e = \int_{\mathcal{C}_p} 1/k \cdot \mathrm{d}s_p \qquad (6.20)$$

If k is essentially a constant over $\mathcal{C}_e$, then

$$d_e \approx 1/k \int_{\mathcal{C}_p} \mathrm{d}s_p \qquad (6.21)$$

If $\mathcal{C}_p$ is a straight line, then

$$d_e \approx 1/k \, [(\delta x)^2 + (\delta y)^2]^{1/2} \qquad \text{from Eq. 6.19}$$

There is often so little variation in k along curves in large-scale maps that the error induced by assuming k is constant becomes negligible. If a project's extent is sufficiently small, it is possible that a single value for k will suffice for the entire map of the project. This can be checked by computing k at the extremes of the project and deciding if the average of these values differs from the extremes too much. If it does, then scale factors can be assigned to baselines individually. The simplest approach is to use the scale factor at the midpoint of the line. If this is unacceptable, the average of the scale factors at the endpoints is better. If this is still unacceptable, use

$$k_{\mathbf{A},\mathbf{B}} = \frac{k_{\mathbf{A}} + 4\,k_{midpoint} + k_{\mathbf{B}}}{6}$$

where $\mathbf{A}$ and $\mathbf{B}$ are the endpoints of the baseline (Stem 1995, p. 50).

How do grid distances compare with slope and normal section distances? Normal section arc lengths are altered by a map projection according to the projection scale factor k, which generally varies over the length of the line. Map projections can be parameterized so that the image of certain curves on the ellipsoid, called **standards**, are arc length true in the projection; standards have everywhere $k = 1$, by definition. Fig. 6.12 on page 124 shows the situation when the endpoints of the line fall between the standards (see sec. 8.3.3 on page 160). The standards appear as solid circles, rather than curves, because the image shows the mapping surface in cross section. $k < 1$ for the entire line, so the length of the projected line is less than that of the normal section. Fig. 6.13 shows the situation when the endpoints fall outside the standards. $k > 1$ for the entire line, so the length of the projected line is greater than that of the normal section. Fig. 6.14 shows the situation when the endpoints fall on different sides of the standards. Part of the line is too big and part is too small.

Distances at elevation are reduced to the ellipsoid by the multiplication of a single number, the elevation factor. Reducing distances from the ellipsoid to a map projection is also the multiplication of a single number (k). The product of the elevation factor and the projection scale factor is a single number that accomplishes both reductions with one multiplication.

6.2. DISTANCE REDUCTIONS

Table 6.3: Coordinates of KERR, ALDRICH, STORRS, and HOLYOKE in various systems. All coordinates refer to the NAD 83(CORS96) reference frame. Linear are in meters, in the Universal Transverse Mercator (UTM), zone 18N; and SPCS83, the State Plane Coordinate System of 1983, zone 0600 (Connecticut).

Station	X	Y	Z
ALDRICH	1 451 508.754	-4 534 331.620	4 230 245.348
KERR	1 451 605.977	-4 534 387.551	4 230 147.979
STORRS	1 450 698.372	-4 534 476.478	4 230 426.808
HOLYOKE	1 414 166.911	-4 515 246.058	4 262 760.044
Geodetic	λ	ϕ	h
ALDRICH	72° 14′ 57″.61678 W	41° 48′ 46″.61736 N	155.622
KERR	72° 14′ 54″.34403 W	41° 48′ 42″.47364 N	152.503
STORRS	72° 15′ 32″.96838 W	41° 48′ 53″.35690 N	195.340
HOLYOKE	72° 36′ 34″.33707 W	42° 12′ 28″.83048 N	21.200
UTM(18N)	x	y	k
ALDRICH	728 479.107	4 632 666.500	1.000 242 424
KERR	728 558.721	4 632 541.110	1.000 242 872
STORRS	727 656.729	4 632 848.294	1.000 237 807
HOLYOKE	697 331.989	4 675 637.989	1.000 079 151
SPCS83(0600)	x	y	k
ALDRICH	346 401.654	261 317.339	0.999 994 9966
KERR	346 477.934	261 189.935	0.999 994 8990
STORRS	345 584.527	261 520.587	0.999 995 1561
HOLYOKE	316 400.613	305 087.718	1.000 052 3970

6.3 Your Turn

The following problems refer to Table 6.3 in which the z-coordinate has been replaced with the projection scale factors for the grid positions. ALDRICH, STORRS, and KERR are stations located on the campus of the University of Connecticut in Storrs. HOLYOKE is in southwestern Massachusetts. The azimuth from KERR to ALDRICH is $329°\,25'\,21''.275\,542$, and the azimuth from ALDRICH to KERR is $149°\,25'\,19''.093\,6287$.

Problem 6.1. Compute the XYZ vector from KERR to ALDRICH. Compute this vector's length.

Problem 6.2. Compute the horizontal distance on the ground from KERR to ALDRICH.

Problem 6.3. Compute the horizontal distance on the ground from ALDRICH to KERR.

Problems 6.4–6.7 illustrate the reduction path one takes to determine a normal section arc distance from a slope distance using a fully rigorous approach.

Problem 6.4. Compute the slope distance from KERR to ALDRICH.

Problem 6.5. Compute the average radius of curvature in the normal section $\overline{R}_\alpha$ for KERR and ALDRICH ($\overline{R}_\alpha = [R_{\alpha_{\text{KERR,ALDRICH}}} + R_{\alpha_{\text{ALDRICH,KERR}}}]/2$). $\alpha_{\text{ALDRICH,KERR}} = 149°\,25'\,19''$, and $\alpha_{\text{KERR,ALDRICH}} = 329°\,25'\,21$.

Problem 6.6. Reduce the slope distance from KERR to ALDRICH to a normal section chord.

Problem 6.7. Reduce the normal section chord from KERR to ALDRICH to a normal section arc.

Problem 6.8. Reduce the normal section from KERR to ALDRICH to a UTM grid distance using the average projection scale factor.

Problem 6.9. Reduce the normal section from KERR to ALDRICH to an SPCS83 grid distance using the average projection scale factor.

Problem 6.10. Determine the UTM grid distance between KERR and ALDRICH by inversing.

Problem 6.11. Determine the SPCS83 grid distance between KERR and ALDRICH by inversing.

Problem 6.12. Repeat problems 6.1–6.11 replacing station ALDRICH with station STORRS. STORRS is about five times farther away than ALDRICH, so the assumptions of the shortcuts become more apparent. The azimuth from KERR to STORRS is $290°\,38'\,32''$, and the azimuth from STORRS to KERR is $110°\,38'\,6''$.

Problem 6.13. Repeat problems 6.1–6.11 replacing station ALDRICH with station HOLYOKE. The azimuth from KERR to HOLYOKE is $325°\,54'\,57''$, and the azimuth from HOLYOKE to KERR is $145°\,40'\,27''$.

Chapter 7

Angles and Point Positioning

One purpose of geometrical geodesy is to create positions with absolute coordinates. Absolute coordinates can be determined by observing angles to celestial objects such as stars. This process determines the position of a single location; the position stands alone, apart from the computed position of any other place. When a position is determined independently, the process is called **point positioning**.

Astronomic point positioning requires observations of a celestial body for which there is an **ephemeris**, which is a catalog of its positions over time. There are ephemerides for all the major celestial bodies such as the Sun, the Moon, the planets, and the many stars. There are also ephemerides for artificial celestial bodies, such as satellites orbiting the Earth. In particular, GNSS satellites are ideally suitable for point positioning, allowing high-accuracy positions to be established even in the most remote places on the Earth. Historically, point positioning was necessary to establish the coordinates of the origins of the old geodetic datums, such as NAD 27, as well as other fiducial stations in the datum.

7.1 North and South

North, also known as **geodetic north**, is the direction along a meridian towards the North Pole. **South** is the direction along a meridian towards the South Pole. Other types of north that arise from physical observations include astronomic north and magnetic north. Astronomic north is also known as true north, and magnetic north is also known as compass north. Still another type of north arises from applying a map projection to a meridian, which usually results in an image of the meridian being a curve that is not parallel to the y-axis of the grid coordinate system. Therefore, the y-axis defines **grid north**.

Azimuths are horizontal angles referenced to north, so each type of north has a corresponding type of azimuth, i.e., a grid azimuth or an astronomic azimuth.

7.1.1 Magnetic north

It has been known for about a century that the Earth's core has two parts, the solid inner core and the liquid outer core (Dubrovinsky and Lin 2009). The liquid outer core is thought to function as an electromagnet that generates the Earth's magnetic field. Magnetic compasses align themselves with these magnetic field lines, which generally are not aligned with meridians.

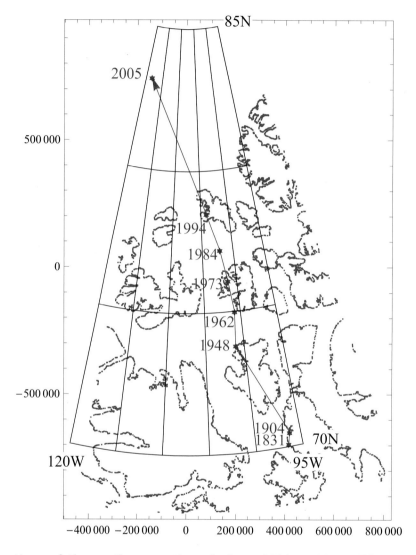

Figure 7.1: Locations of the north magnetic pole from 1831 to 2005. The polar stereographic projection has a central scale factor of 1.0, using GRS 80 as the reference surface (data courtesy of NASA).

7.1. NORTH AND SOUTH 131

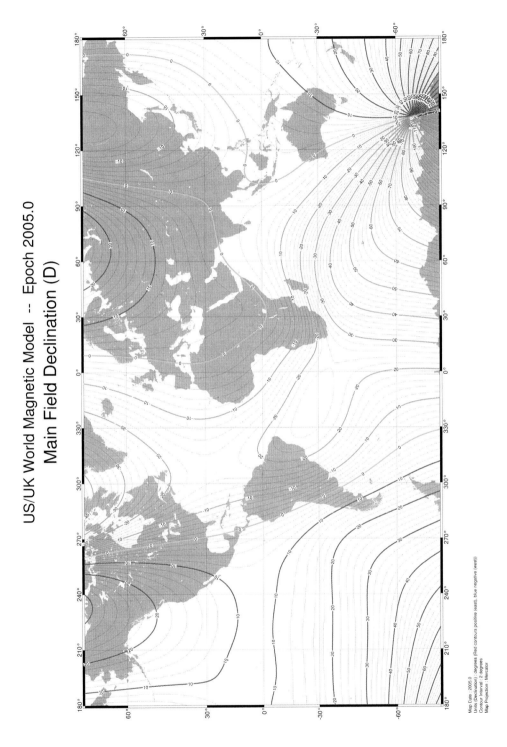

Figure 7.2: Magnetic declination in 2005 as computed by the NGA World Magnetic Model (courtesy of NGA). Positive declinations indicate that magnetic north is west of geodetic north, and negative declinations indicate that magnetic north is east of geodetic north.

The magnetic poles are not located at the geodetic poles. As of 2005.0 the magnetic North Pole was located at 118.32° W, 83.21° N, which is about 760 km from the geodetic North Pole. The magnetic South Pole was at 137.86° E, 64.53° S, which is about 2840 km from the geodetic South Pole. The magnetic poles move and even change polarity (Glazmaier and Roberts 1995). From 1831 to 2005, the magnetic North Pole's velocity varied from 0.6 km/year to 52.7 km/year with an average of 9 km/year (Fig. 7.1).

The angular difference in azimuths between geodetic north and magnetic north is called **magnetic declination**. In particular, $\alpha_m = \alpha - \delta_m$, where α_m is magnetic azimuth, α is geodetic azimuth, and δ_m is magnetic declination. δ_m can be positive or negative. Various scientific and governmental entities model the Earth's magnetic field. The United States National Geospatial-Intelligence Agency (NGA) produces one such model, the World Magnetic Model (WMM). The U.S. National Geophysical Data Center (NGDC) and the British Geological Survey (BGS) created the WMM (McLean et al. 2004). Figure 7.2 on the previous page is an output of WMM showing the magnetic declination of the Earth's north magnetic pole in 2005.

Magnetic storms can cause compass directions to shift by 10° or more over a few hours, so magnetic north is often not a reliable geodetic azimuth datum even if magnetic declination has been accounted for (Campbell 1997; Blakely 1995; Sleep and Fujita 1997).

7.1.2 Astronomic north

Polaris, the North Star, happens to be currently located very nearly at the zenith of the Earth's rotational axis. Polaris, therefore, has no apparent motion to the unaided eye due to the Earth's diurnal rotation, as if it were at a fixed place in the night sky of the Northern Hemisphere. This gives Polaris a distinguished place in the history of navigation because it became a natural reference for latitude and an easy way to determine the direction of north. The steadfastness of Polaris earned astronomic north the honorific of "true north," but this turns out to be something of a misnomer because the Earth's polar motion continuously changes astronomic north (see section 7.1.2 on page 136).

With a theodolite and a clear view of Polaris, one can observe the **astronomic latitude** Φ of the observation station: the angle between the local horizontal and the line-of-sight to Polaris is astronomic latitude. **Astronomic longitude** Λ can be deduced, for example, by knowing the temporal offset from the time in Greenwich, England (Bomford 1962, p. 281).

Astronomic coordinates

Astronomic point positioning requires a new category of coordinate systems, those whose orientation depends on the direction that gravity acts at the coordinate system's origin. When a telescope is used to measure, say, a zenith angle to a star, the telescope's platform must be level in order for the angle to be observed correctly. An instrument is leveled by adjusting the orientation of the platform, often using fine-thread precision machine screws, until bubble levels on the platform indicate it is level. One could conceptualize this horizontal platform as being perpendicular to the direction a plumb bob would hang if suspended directly below the platform. A plumb bob hangs so as to follow the direction of the force of gravity at that location.

Bomford (1962, p. 83) writes that astronomical observations "...are of great interest and value, but they do not constitute a useful coordinate system." But why not? Suppose a theodolite is set up and leveled perfectly on flat terrain. This theodolite has a plumb bob hanging below it perpendicularly to the horizontal stage of the telescope. An operator uses the theodolite to observe Polaris and records, without error, the astronomic latitude. Suddenly, a cataclysmic earthquake thrusts up a tall mountain immediately to the south of the instrument without either harming or even disturbing it. The operator inspects the instrument and sees it is no longer in level; the plumb bob no longer hangs perpendicularly to the instrument stage even though the instrument was neither harmed nor disturbed. The mountain now occupies a great volume of space that just before was occupied by air. The material of the mountain is vastly more dense than air, creating a gravitational pull towards it, pulling the plumb bob away from its previous orientation. Had the mountain been there all along, the operator would have leveled the instrument according to the deflected direction of the plumb line and would have read a different latitude, even though the the readings were without error. Astronomic latitude is directly impacted by the Earth's gravity field, which changes in magnitude and direction from place to place due to inhomogeneous mass distributions in the Earth. Therefore, astronomic positions change as the Earth's shape changes, even excluding real motions due to plate tectonics, erosion, mass wasting, and so on.

Deflection of the vertical

The vertical is usually not parallel to the normal, meaning the direction gravity acts at a location is not the direction perpendicular to a reference ellipsoid. Therefore, coordinate systems based on physical measurements differ from the geometric coordinate systems in which we need to perform computations. The angle between the local direction of gravity and the ellipsoidal normal is called the **deflection of the vertical**. The largest deflections occur near islands surrounded by deep waters: deflections modeled by the Earth Gravitational Model 2008 (EGM2008) near Hawaii and Guam exceed 100 arc seconds (Pavlis et al. 2008).

The deflection of the vertical causes angular traverse loop misclosures, as do instrument setup errors, the Earth's curvature, and environmental factors introducing errors into measurements. The practical consequence of the deflection of the vertical is that observed angles differ from the angles that result from the pure geometry of the stations referred to reference ellipsoids. It is as if the observing instrument were misleveled, producing traverses that do not close. For example, Shalowitz (1938, p. 13, 14) reported that deflections of the vertical created discrepancies between astronomic coordinates and geodetic (computed) coordinates up to a minute of latitude in Wyoming. Control networks for large regions cannot ignore these discrepancies, and remain geometrically consistent, especially in and around regions of great topographic relief.

The deflection of the vertical is a single spatial angle, but it can be decomposed into two components, one (ξ) in the direction of the meridian (north-south) and one (η) in the direction of the prime vertical (east-west). The relationships among geodetic coordinates, astronomic coordinates, and the deflection of the vertical are derived using spherical trigonometry (see section 7.2.1) (Jekeli 2005, p. 2-57–61):

$$\xi = \Phi - \phi \tag{7.1}$$

$$\eta = (\Lambda - \lambda) \cos \phi \tag{7.2}$$

$$\eta = (A - \alpha) \cot \phi \tag{7.3}$$

where A and α are astronomic and geodetic azimuths, respectively, and all other quantities have their usual meanings. The **(extended) Laplace condition** is

$$A - \alpha = (\Lambda - \lambda) \sin \phi + (\xi \sin \alpha - \eta \cos \alpha) \cot \zeta' \tag{7.4}$$

where ζ' is the observed zenith angle to an observed object. Deflections are the objects of interest for certain navigational and geophysical purposes. If the geodetic coordinates of observation and orientation stations are known, Eq. 7.1, 7.2, and 7.3 can be used to determine high-accuracy deflections (better than $1''$). For terrestrial surveying, ζ' is usually close to $90°$, so the second term on the right-hand side of Eq. 7.4 is nearly zero and often omitted, resulting in the **Laplace condition** (also known as **Laplace's equation**):

$$A - \alpha = (\Lambda - \lambda)\sin\phi \tag{7.5}$$

Rearranging Eq. 7.2 and substituting for $(\Lambda - \lambda)$ in Eq. 7.5 results in

$$\alpha = A - \eta \tan\phi \tag{7.6}$$

The term $\eta \tan\phi$ is called the **Laplace correction**, which can be used transform an astronomic azimuth A to a geodetic azimuth α by accounting for the deflection of the vertical.

Example 7.1. The NGS data sheet puts station Y 88 (PID LX3030) at $\lambda = 72°\,15'\,02''.042\,45$ W, $\phi = 41°\,48'\,44''.784\,64$ N, and $h = 157.280$ m. NGS data sheets do not, unfortunately, report the deflection of the vertical. However, the NGS utility DEFLEC99 computes them (NGS 2001). From DEFLEC99, $\eta = 2''.66$ and $\xi = -6''.53$ at Y 88. Compute the Laplace correction.
Solution: $\eta \tan\phi = 2''.66 \tan 41.812\,440\,18° = 2''.38$
$\square$

The Laplace correction is reported on NGS data sheets. For Y 88 it is

```
LX3030   LAPLACE CORR-           -2.38   (seconds)                    DEFLEC99
```

Notice that the algebraic sign of the NGS value is opposite that of the quantity computed from Eq. 7.6. The NGS value needs to be added to an astronomic azimuth, rather than subtracted, so care must be taken when using the NGS value.

Deflection of the vertical and directions

The deflection of the vertical affects directions read from the horizontal circle of leveled instruments such as total stations. Geodetic azimuth α equals observed azimuth α' plus an offset $\delta\alpha$ due to the deflection of the vertical: $\alpha = \alpha' + \delta\alpha$. For a total station at **T** pointed at **S** (Elithorp Jr. and Findorff 2003, p. 176)

$$\delta\alpha_\mathbf{S} = -(\xi \sin\alpha_\mathbf{S} - \eta \cos\alpha_\mathbf{S})\cot\zeta_\mathbf{S} \tag{7.7}$$

where $\alpha_\mathbf{S}$ is geodetic azimuth from **T** to **S**, and $\zeta_\mathbf{S}$ is the zenith angle from **T** to **S** (mark-to-mark). Recall that bearings are determined from the difference of pointings to a foresight and a backsight. For a total station with a backsight station **B** and a foresight station **F**, Eq. 7.7 is used to determine the bearing $\beta_{\mathbf{B},\mathbf{F}}$ as (Elithorp Jr. and Findorff 2003, p. 186)

$$\begin{aligned}\beta_{\mathbf{B},\mathbf{F}} &= \alpha_\mathbf{F} - \alpha_\mathbf{B} \\ &= (\alpha'_\mathbf{F} + \delta\alpha_\mathbf{F}) - (\alpha'_\mathbf{B} + \delta\alpha_\mathbf{B}) \\ &= \alpha'_\mathbf{F} - \alpha'_\mathbf{B} + \delta\alpha_\mathbf{F} - \delta\alpha_\mathbf{B}\end{aligned} \tag{7.8}$$

α' is a reading of a horizontal circle and need not be an azimuth (in fact, it need not have any particular relationship to geodetic north), but $\alpha_\mathbf{S}$ in Eq. 7.7 is a geodetic azimuth and must be

7.1. NORTH AND SOUTH

Table 7.1: Observed horizontal circle readings and zenith angles from Y 88 to ALDRICH and CANR01.

Station	α'	ζ
ALDRICH	109° 11′ 15″	90° 48′ 50″
CANR01	172° 17′ 00″	91° 56′ 10″

determined as such. If an angle is turned to the forward station with the circle having been zeroed on the backsight, then $\alpha'_{\mathbf{B}} = 0$ and we have

$$\beta_{\mathbf{B},\mathbf{F}} = \alpha'_{\mathbf{F}} + \delta\alpha_{\mathbf{F}} - \delta\alpha_{\mathbf{B}} \tag{7.9}$$

The correction for the deflection of the vertical is usually quite small except for observing stars at high angles above the horizon. Notably, horizontal sightings are unaffected by Eq. 7.7 entirely. However, the correction is not always small, even for terrestrial surveying, and should be checked before a decision to ignore it is made.

Example 7.2. Observed directions and zenith distances to ALDRICH (A) and CANR01 (C) observed at Y 88 are given in Table 7.1. The geodetic azimuth from Y 88 to ALDRICH is $\alpha_{\text{ALDRICH}} = 61° 2′ 12″.344\,309 = 61.036\,76°$. With $\xi = -6″.53$ and $\eta = 2″.66$ at Y 88, what is $\beta_{\mathbf{A},\mathbf{C}}$?

Solution: Applying Eq. 7.7 at ALDRICH gives

$$\begin{aligned}
\delta\alpha_{\mathbf{A}} &= -(\xi \sin\alpha_{\mathbf{A}} - \eta \cos\alpha_{\mathbf{A}})\cot\zeta_{\mathbf{A}} \\
&= -(-6.53″ \sin 61.036\,76° - 2.66″ \cos 61.036\,76°)\cot 90.813\,87° \\
&= 0″.099\,4595
\end{aligned}$$

To apply Eq. 7.7 at CANR01 we need $\alpha_{\mathbf{C}}$, which is unknown, but can be easily approximated by $\alpha_{\mathbf{C}} \approx \alpha_{\mathbf{A}} + \alpha'_{\mathbf{C}} - \alpha'_{\mathbf{A}} = 61.036\,76° + 172.2833° - 109.1875° = 124.1326°$. Applying Eq. 7.7 at CANR01 gives

$$\begin{aligned}
\delta\alpha_{\mathbf{C}} &= -(\xi \sin\alpha_{\mathbf{C}} - \eta \cos\alpha_{\mathbf{C}})\cot\zeta_{\mathbf{C}} \\
&= -(-6.53″ \sin 124.1326° - 2.66″ \cos 124.1326°)\cot 91.9361° \\
&= -0″.132\,263
\end{aligned}$$

Then Eq. 7.8 gives

$$\begin{aligned}
\beta_{\mathbf{A},\mathbf{C}} &= \alpha'_{\mathbf{C}} - \alpha'_{\mathbf{A}} + \delta\alpha_{\mathbf{C}} - \delta\alpha_{\mathbf{A}} \\
&= 172.2833° - 109.1875° - 0″.132\,263 - 0″.099\,4595 \\
&= 63° 5′ 45″
\end{aligned}$$

This is the same answer, to the nearest arc second, that is obtained without the corrections.
□

Precession and nutation

Another set of problems arises when Polaris is used as the reference for astronomic latitude because (i) Polaris is not *exactly* at the zenith of the Earth's rotational axis and (ii) the Earth's rotational axis slowly describes a generally circular path on the celestial sphere around the average zenith through processes called **precession** and **nutation**.

Precession causes the Earth's rotational axis to slowly trace a circle on the celestial sphere, a motion resembling that of a spinning top. The Earth's precession is caused by the equatorial bulges not aligning in the **plane of the ecliptic** (the plane in which the Earth orbits the Sun), thereby giving rise to a torque from the gravitational attraction of the Sun (Vaníček and Krakiwsky 1996, p. 59). The earth's precession is slow, with its axis returning to a previous orientation once in approximately 25 765 years, a period known as a **Platonic year**. Likewise, the equatorial bulges are not aligned with the Moon's orbital plane, which is inclined 5° 11′ to the ecliptic. The intersection of the Moon's orbital plane with the ecliptic is known as the **nodal line**, and the nodal line rotates once in 18.6 years, the **Metonic cycle**. This constant realignment of the Moon with the Earth also affects the orientation of the Earth's rotational axis, causing a motion called **nutation** (Vaníček and Krakiwsky 1996; Volgyesi 2006).

Coordinate systems ought not change haphazardly due to diverse effects in the physical universe. Geodetic coordinate systems are geometrical, with the immutability of mathematics. Were this not so, positions would change over time in the absence of real physical motion of the stations, which is highly undesirable. The motions of the Earth's rotational axis are continuously monitored by International Earth Rotation and Reference Systems Service (IERS) and cataloged as **Earth rotation parameters**. Corrections computed from the Earth rotation parameters allow astronomic observations to be used for geodetic point positioning and GNSS orbits to be computed in a reference frame free from these motions. Nonetheless, an empirical component of geodetic positions always remains, and developing highly stable reference frames is very challenging.

7.1.3 Geodetic north

As we can see, orienting a survey to geodetic north cannot be done in the field simply by pointing an instrument at some naturally occurring object, like the North Star. Geodetic north is not observable; it is not a physical quantity. Furthermore, geodetic north generally differs among datums at any given place, since different datums place and orient their meridians differently.

Geodetic north can be inferred from geodetic coordinates by inversing, which computes the forward azimuth of the geodesic segment between two stations whose positions are given in geodetic coordinates (see section 6.1.8 on page 116). For example, suppose a POB and a backsight's coordinates were determined using GNSS positioning. These coordinates could be used as inputs to the geodetic inverse problem. The geodetic orientation of this line can be used as the directional basis of the survey and propagated throughout the positions by traversing. As an alternative to inversing, surveyors often begin a survey on a monument with published geodetic coordinates that is within sight of another station, also with published coordinates. If the other station is an azimuth mark for the observation station, then the azimuth from the monument to the azimuth station is also published, which saves the trouble of doing the computation. The following is an excerpt from the NGS data sheet for station STORRS in Connecticut, which has a primary azimuth-mark STORRS AZ MK 2:

```
LX4976:                  Primary Azimuth Mark                  Grid Az
LX4976:SPC CT      -  STORRS AZ MK 2                         030 40 22.4
LX4976:UTM  18     -  STORRS AZ MK 2                         029 10 12.6
LX4976
LX4976|---------------------------------------------------------------|
LX4976| PID    Reference Object                  Distance    Geod. Az |
LX4976|                                                      dddmmss.s|
LX4976| LX4985 STORRS AZ MK 2                    APPROX. 0.6 KM 0305954.0 |
```

The Z-axes of conventional terrestrial reference frames are set parallel to the direction of the Earth's rotational axis averaged over some period of time. The IERS publishes Conventions (McCarthy and Petit 2004) that put forth the relationships between the reference systems and the conventional reference pole.

Creating regional horizontal geodetic datums required astronomic point positioning to establish the datum's origin and orientation, as well as to control error propagation in the triangulation network. However, such point positioning is very time-consuming and difficult. Thus arises the need for geodetic surveying to produce absolute geodetic coordinates using relative surveying practices of observing angles and distances to terrestrial stations rather than to celestial bodies. This requires the use of spherical trigonometry.

7.2 Spherical Trigonometry

As with plane surfaces, angles are used to solve the direct and inverse problems on spherical reference surfaces. Spherical trigonometry underpins geodetic surveying, which is done on a reference ellipsoid using modifications to the spherical formulæ.

A plane angle is the rotational offset between two straight lines that intersect at a single point. However, a sphere has no straight lines, so we need another type of angle for geodetic surveying. Great circles are the equivalents of straight lines on spheres. Figure 7.3 on the next page shows two great circles in blue and red as the intersections of the two vertical planes with the sphere. Suppose the great circles intersect at any point **A** on the sphere. The sphere has a tangent plane at **A** and the projections of the great circles onto that tangent plane are straight-line segments (shown in their respective colors). The angle between those projected segments is a **dihedral angle**. Dihedral angles can also be defined as the rotational offset between two planes that intersect in a line. Dihedral angles on a sphere are called **spherical angles**.

The basic relationship between an angle and the length of a circular arc segment ($d = r\,\theta$) is the key concept in spherical trigonometry because it provides a simple duality between angles and distances on the unit sphere: if $r = 1$, then $d = \theta$. Figure 7.4 on page 139 depicts a **spherical triangle** on a unit sphere with center **O** (Weisstein 2008). Points **A**, **B**, and **C** are connected by three great-circle arc segments of lengths a, b, c, with side names corresponding to opposite dihedral angles, as is customary. α, β, and γ are spherical angles. Planar angles a, b, and c, which are also the planes of the great circle segments, are formed by radii to the points. The lengths of the triangle sides are denoted exactly the same as the three planar angles because, by $d = r\,\theta$ and $r = 1$, they have the same values.

Figure 7.3: Dihedral angle.

7.2.1 Spherical law of sines

The (planar) laws of sines and cosines have equivalents for spherical trigonometry (see appendix B). Referring to Fig. 7.4, the **spherical law of sines** can be written

$$\frac{\sin \alpha}{\sin a} = \frac{\sin \beta}{\sin b} = \frac{\sin \gamma}{\sin c} \tag{7.10}$$

The above equation has exactly the same structure as the planar law of sines except the denominators in Eq. 7.10 are sines of distances rather than distances, which are the denominators in planar equations. However, these distances equal the angles with the same names, so treating them as angles is sensible.

7.2.2 Spherical law of cosines

The following equations are called the **cosine rule for sides**:

$$\cos a = \cos b \cos c + \sin b \sin c \cos \alpha \tag{7.11}$$
$$\cos b = \cos a \cos c + \sin a \sin c \cos \beta \tag{7.12}$$
$$\cos c = \cos a \cos b + \sin a \sin b \cos \gamma \tag{7.13}$$

and the **spherical law of cosines** has the form:

$$\cos \alpha = -\cos \beta \cos \gamma + \sin \beta \sin \gamma \cos a \tag{7.14}$$
$$\cos \beta = -\cos \alpha \cos \gamma + \sin \alpha \sin \gamma \cos b \tag{7.15}$$
$$\cos \gamma = -\cos \alpha \cos \beta + \sin \alpha \sin \beta \cos c \tag{7.16}$$

7.3. POSITIONING ON A SPHERE

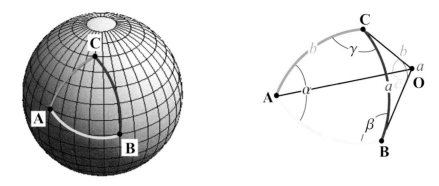

Figure 7.4: Triangle on a sphere (courtesy of Eric W. Weisstein).

These equations bear less resemblance to their planar equivalents, but a resemblance nevertheless remains. The structure of the planar cosine law is that the cosine of the angle is framed by two distances. The third distance appears in the equation as the only negative term in the numerator and is missing in the denominator. A similar pattern can be seen in the spherical versions.

7.3 Positioning on a Sphere

Assuming a single radius to represent the entire Earth and using spherical trigonometry for positioning is not accurate enough for geodetic purposes. Nonetheless, seeing how positioning with spherical trigonometric formulæ is not so different from positioning with their planar cousins can help one understand how the process is carried out on a reference ellipsoid. (Spherical trigonometry techniques do occur in geodesy for other purposes, such as developing equations for astronomic positioning.)

Let's define **colatitude** θ to be the angle in the meridional plane between the Z-axis and the point of interest (Fig. 7.5 on the next page). Colatitude is related to geodetic latitude by $\phi = \pi/2 - \theta$.

Figure 7.5 is a redrawn version of Fig. 7.4 that places **C** at the North Pole with the angles labeled using geodetic nomenclature. γ is equal to $\delta\lambda$, the change in longitude between the observation station **A** and the observed station **B**; angle b is replaced with $\theta_\mathbf{A}$, and angle a is replaced with $\theta_\mathbf{B}$. This nomenclature is somewhat backwards from that in Fig. 7.4, but is necessary because $\theta_\mathbf{A}$ is the colatitude of **A** and $\theta_\mathbf{B}$ is the colatitude of **B**. α is the spherical angle from the meridional plane containing **A** to the great circle plane c between **A** and **B**; this is the definition of the azimuth from **A** and **B**. β is the spherical angle from the meridional plane containing **B** to the great circle plane between **A** and **B**; this is the definition of the negative of the azimuth from **B** and **A**. a and b are meridional distances from the North Pole and are equal to $b = R\theta_\mathbf{B}$ and $a = R\theta_\mathbf{A}$, where R is the radius of the sphere, which will be taken as identically 1 (one) hereafter. c is also the observed

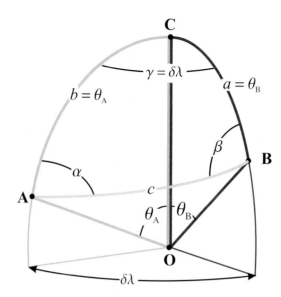

Figure 7.5: Spherical triangle that is the basis of geodetic surveying. The triangle resides on a unit sphere with center at **O** and North Pole at **C**. **A** is the observation station assumed to be known. **B** is the observed station. The observed quantities are α and c.

great circle distance between **A** and **B**.

7.3.1 The direct problem on the sphere

The direct problem on the sphere is as follows: given point **A**, determine point **B** at distance c from **A** on the great circle oriented at azimuth α. Colatitude $\theta_{\mathbf{B}}$ comes directly from Eq. B.2:

$$\cos\theta_{\mathbf{B}} = \cos\theta_{\mathbf{A}} \cos c + \sin\theta_{\mathbf{A}} \sin c \cos\alpha \qquad (7.17)$$

where we have used the duality between spherical distance c and its corresponding angle. Two checks can be performed. First, if $c = 0$, then **A** = **B**.

$$\begin{aligned}
\cos\theta_{\mathbf{B}} &= \cos\theta_{\mathbf{A}} \cos c + \sin\theta_{\mathbf{A}} \sin c \cos\alpha \\
&= \cos\theta_{\mathbf{A}} \cos 0 + \sin\theta_{\mathbf{A}} \sin 0 \cos\alpha \\
&= \cos\theta_{\mathbf{A}}
\end{aligned}$$

Second, if $\alpha = 0$, then **B** is due north of **A**, and if $c = \theta_{\mathbf{A}}$, then **B** = **C**, the North Pole:

$$\begin{aligned}
\cos\theta_{\mathbf{B}} &= \cos\theta_{\mathbf{A}} \cos\theta_{\mathbf{A}} + \sin\theta_{\mathbf{A}} \sin\theta_{\mathbf{A}} \cos 0 \\
&= \cos\theta_{\mathbf{A}} \cos\theta_{\mathbf{A}} + \sin\theta_{\mathbf{A}} \sin\theta_{\mathbf{A}} \\
&= \cos^2\theta_{\mathbf{A}} + \sin^2\theta_{\mathbf{A}} \\
&= 1
\end{aligned}$$

So $\theta_{\mathbf{B}} = 0$, the value for the North Pole, as expected.

With θ_B from Eq. 7.17, $\delta\lambda$ is found from the spherical sine law ($\sin\alpha/\sin a = \sin\delta\lambda/\sin c$):

$$\frac{\sin\delta\lambda}{\sin c} = \frac{\sin\alpha}{\sin\theta_B}$$
$$\sin\delta\lambda = \frac{\sin\alpha\,\sin c}{\sin\theta_B} \qquad (7.18)$$

where again we have used the duality between the spherical distance a and its corresponding angle θ_B. Applying the first check as above, if $c = 0$, then the right-hand side of Eq. 7.18 is zero, so $\delta\lambda = 0$, as expected. The second check results in an indetermine longitude value, which is correct: the North Pole has no determinate longitude. However, suppose $\alpha = 0$ so that **B** is on the same meridian as **A**. Then we have

$$\frac{\sin\delta\lambda}{\sin c} = \frac{\sin 0}{\sin\theta_B}$$
$$\sin\delta\lambda = 0$$
$$\delta\lambda = 0$$

7.3.2 The inverse problem on the sphere

The inverse problem on the sphere is as follows: given **A** and **B**, determine c and α. From this problem statement we see that $\delta\lambda = \lambda_B - \lambda_A$, $\theta_A = \pi - \phi_A$, and $\theta_B = \pi - \phi_B$. Spherical distance d comes directly from $d = Rc$, where c is determined from Eq. 7.13 as

$$\cos c = \cos\theta_A\,\cos\theta_B + \sin\theta_A\,\sin\theta_B\,\cos\delta\lambda$$

Therefore,
$$d = R\arccos(\cos\theta_A\,\cos\theta_B + \sin\theta_A\,\sin\theta_B\,\cos\delta\lambda) \qquad (7.19)$$

Notice that Eq. 7.19 is the same as the formula for great circle distance (Eq. 6.6). Azimuth α comes directly from the spherical sine law (Eq. 7.10):

$$\sin\alpha = \frac{\sin\theta_B\,\sin\delta\lambda}{\sin c} \qquad (7.20)$$

although care must be taken to disambiguate the quadrant because $0 \leq \alpha < 2\pi$ (see section 8.4 on page 160 for an example).

For full mathematical rigor, the above formulæ must be scrutinized for singularities that arise for special cases. For example, the right-hand side of Eq. 7.20 is not defined if $c = 0$, but this is sensible because $c = 0$ when the length of the observed line is zero, which implies that the observed point is coincident with the observation station.

7.4 Grid Angles

Mapping in a grid coordinate system involves two additional horizontal angles: convergence and the arc-to-chord correction.

7.4.1 Convergence

Map projections that do not change the angle at which intersecting lines meet are called **conformal**. Map projections can, and do, change the orientation of intersecting lines. Therefore, the projected image of a meridian and a parallel under a conformal projection will meet at a right angle, but the meridian will not necessarily be parallel with the y-axis nor will the parallel necessarily be parallel with the x-axis. So, if a mapper were to use a geodetic azimuth from a POB to the survey's orienting mark but compiled the map using grid coordinates, then the survey would be misaligned with respect to the grid. If a map is to be compiled in a map projection, then the geodetic azimuths between stations need to be changed to their projected equivalents. The difference between geodetic and grid azimuth is called **grid declination** or **convergence**. Convergence is positive when grid north is east of geodetic north.

Convergence can be obtained in several ways. NGS data sheets give convergence for both the UTM and the SPCS grids. Here is an excerpt from the Y 88 data sheet:

```
LX3030;                    North         East      Units  Scale Factor  Converg.
LX3030;SPC  CT     -     261,260.205   346,299.832  MT    0.99999495    +0 19 52.2
LX3030;UTM  18     -   4,632,606.701   728,378.794  MT    1.00024186    +1 50 01.8
```

Coordinate calculator programs, such as CORPSCON, also give convergence. Convergence γ is added to grid azimuth to obtain geodetic azimuth (Fig. 7.6 on the next page):

$$\alpha = \alpha_g + \gamma \tag{7.21}$$

Convergence can change significantly even over short distances. For example, the SPCS83 convergence is $0° 19' 55''.1$ at station ALDRICH, which is 116.765 m (geodesic) from Y 88. This is about a 3-arc second difference from the convergence at Y 88.

7.4.2 Arc-to-chord correction

Figure 7.6 shows the image of a meridian and a normal section arc as they appear in a mapping grid. An observation station is at **O**, and a forward station is at **F**. The image of the projected geodesic from **O** to **F** is the curve next to the straight-line chord connecting them (the projected image of a geodesic is usually curved). The grid azimuth of the geodesic is α_n, whereas the grid azimuth of the chord, which would be computed by inversing between the grid coordinates of the stations, is α_g. The geodetic azimuth of the geodesic is $\alpha = \alpha_n + \gamma$. The angle δ formed from the geodesic to its chord is called the **arc-to-chord correction**, so $\alpha_g = \alpha_n + \delta$. In the figure, δ is a negative angle. The grid azimuth between two stations is related to geodetic azimuth, convergence, and the arc-to-chord correction by

$$\alpha_g = \alpha - \gamma + \delta$$

and the geodetic azimuth can be obtained from grid azimuth by $\alpha = \alpha_g + \gamma - \delta$.

The arc-to-chord correction is exceedingly small for lines shorter than continental and can be safely ignored for all but highly rigorous work. In fact, there other, even smaller angles that arise between normal sections, geodesics, and their projected images (Rapp 1989a).

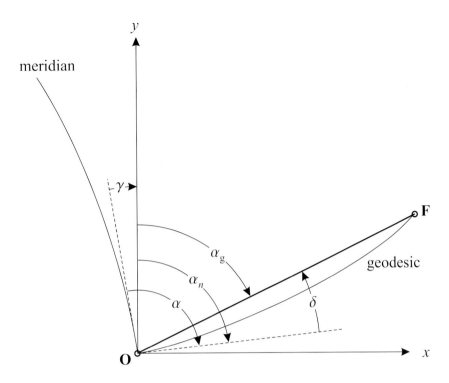

Figure 7.6: The image of a meridian under a map projection is generally not parallel to the grid y-axis. Similarly, the image of a normal section is not a chord.

7.5 Your Turn

Problem 7.1. If you're standing at the North Pole looking at the horizon, in what direction are you facing?

Problem 7.2. If you're standing at the South Pole looking at the horizon, in what direction are you facing?

Problem 7.3. If you're at the center of the Earth, in what direction are you facing?

Problem 7.4. For problems 7.4 to 7.7, see Table 6.3 for coordinates of stations ALDRICH and KERR. Use DEFLEC99 to find η and ξ at ALDRICH, and compute the Laplace correction.

Problem 7.5. Compute the mark-to-mark zenith angle $\zeta_{Y,K}$ from Y 88 to KERR in DMS. (Hint: use KERR's coordinates in Y 88's *ENU* system.)

Problem 7.6. Compute the mark-to-mark zenith angle $\zeta_{K,Y}$ from KERR to Y 88 in DMS. What is $\zeta_{Y,K} + \zeta_{K,Y}$ in DMS? Why does the sum not equal $180°\,0'\,0''$? Would you get the same sum for a station the same distance from Y 88 as KERR but at a different azimuth?

Problem 7.7. A total station is set up at Y 88. The observed direction from Y 88 to ALDRICH is $\alpha'_A = 109°\,11'\,15''$. Observed direction from Y 88 to KERR is $\alpha'_K = 160°\,0'\,45''$. Using values given at Y 88, the computed zenith angles, compute $\beta_{A,K}$ after correcting for the deflection of the vertical.

Problem 7.8. CT SPCS83 coordinates for Y 88 are (346 299.822, 261 260.207), and the CT SPCS83 coordinates for ALDRICH are (346 401.6542, 261 317.339) in meters. Compute the grid azimuth from Y 88 to ALDRICH, and verify that the geodetic azimuth is 61° 2′ 13″.

Chapter 8

Map Projections

Map projections transform coordinates between a geographic coordinate system (λ, ϕ) and a Cartesian coordinate system (x, y), either by graphical or mathematical means. **Forward** mappings transform coordinates from geographic systems to Cartesian systems. Height is not included. Projected coordinates can be augmented by a third dimension to represent height, but map projection formulæ make no use of a height coordinate. The projected image of a grid of parallels and meridians is called a **graticule**. A system of lines parallel with the x- and y-axes form a map projection **grid**, which is always rectangular. **Inverse** mappings transform coordinates from Cartesian systems to geographic systems. No information is lost, or gained, in a map projection. This is different than the use of the word "projection" in the broader mathematical context, in which a projection is not invertible and projected points cannot be taken back to the original.

8.1 Developable Surfaces

Maps are compiled onto flat surfaces, such as pieces of paper or computer screens. A surface is topologically planar if it can be made flat by deformations such as twistings and stretchings, and a cut or two is allowed. For example, cylinders are topologically planar because they can be cut up the side and pressed flat, whereas a sphere, or even a hemisphere, cannot. A topologically planar surface is called **developable**.

Ancient map projections were geometric constructions, but most modern projections are not: they have no geometric construction and use developable surfaces that have no name. Ancient map projections mapped spheres to cones, cylinders, and planes. These surfaces have become erroneously entrenched as being the only surfaces that are developed into maps, and map projections are widely classified as being cylindrical, planar, or conic.

8.1.1 Cylindrical map projections

Suppose a cylinder of radius R is wrapped around a sphere, also of radius R, such that the cylinder touches the sphere at its equator and the cylinder's axis coincides with the sphere's axis. Define a map projection

$$(x, y) = f(\lambda, \phi) = (R[\lambda - \lambda_{origin}],\ R \tan \phi) \tag{8.1}$$

λ_{origin} is called the **central meridian** of the projection, which is a projection parameter whose value is chosen to place some longitude of interest at an easting of zero. Latitudes are mapped to

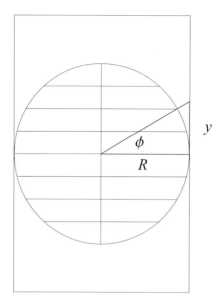

Figure 8.1: A straightforward approach to mapping latitude onto a cylinder. The northing is $y = R\tan\phi$.

northings by $y = R\tan\phi$ (see Fig. 8.1). Longitudes are mapped to eastings by $x = R(\lambda - \lambda_{origin})$, with $-\pi \leq \lambda < \pi$. This is the mathematics behind a projection that can be visualized as a light bulb being at the center of a transparent sphere shining out in all directions and casting shadows of the points of interest onto a cylindrical sheet wrapped around the sphere. Figure 8.2 shows country boundaries on the sphere in red and their projection on the cylinder in blue. Cutting the cylinder along the international date line produces the map in Fig. 8.3.

Equation 8.1 projects the equator arc length true, but no other parallel or meridian. Equation 8.1 can be augmented so that the image of any chosen parallel is arc length true:

$$(x,y) = f(\lambda, \phi) = (R[\lambda - \lambda_{origin}]\cos\phi_1, R\tan\phi) \tag{8.2}$$

ϕ_1 is called the **standard parallel** of this projection.

The following formula is another cylindrical map projection that takes points on a sphere of radius R to points on a cylinder, also of radius R, such that the cylinder's axis coincides with the sphere's axis, and the cylinder touches the sphere at its equator:

$$(x,y) = f(\lambda, \phi) = (R[\lambda - \lambda_{origin}]\cos\phi_1, \ R\phi) \tag{8.3}$$

Snyder (1987, p. 90) calls Eq. 8.3 the **Equidistant Cylindrical Projection** because it projects the standard parallel and all the meridians arc length true. The Equidistant Cylindrical Projection is arguably the simplest projection: Eq. 8.3 becomes a unity mapping for $R = 1$, $\phi_1 = 0$, and $\lambda_{origin} = 0$: $(x, y) = f(\lambda, \phi) = (\lambda, \ \phi)$. When the equator is the standard parallel, Eq. 8.3 is often called the Plate Carrée projection or the Simple Cylindrical Projection. The graticule of the spherical Simple Cylindrical Projection is a square grid (see Fig. 8.4), which can be very helpful for displaying remotely sensed images whose pixels are squares. Some GIS offer a "no projection" option for geographic coordinates, which usually means treating longitude (in decimal degrees) as

8.1. DEVELOPABLE SURFACES

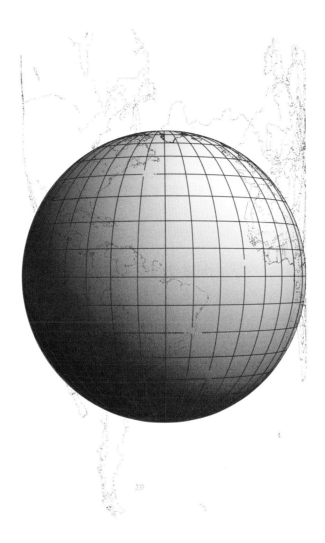

Figure 8.2: Projection of political boundaries onto a cylinder using $(x, y) = f(\lambda, \phi) = (R\lambda, R\tan\phi)$. Points in red are on the sphere. Points in blue are their projection onto the cylinder.

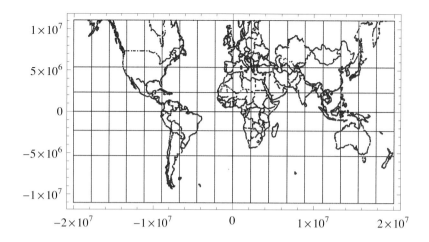

Figure 8.3: Map produced by cutting the cylinder in Fig. 8.2 along the international date line with $R = 6\,378\,137$ m.

the x-coordinate and latitude (in decimal degrees) as the y-coordinate. This is a nonmetric[1] version of the Simple Cylindrical Projection because longitude and latitude are not in radians.

The maps in Fig. 8.3 and Fig. 8.4 differ significantly. For example, the shapes of the countries are different, the scales of the y-axes are different, and the amount of the world that can be shown is different. Clearly, a developable surface can be projected in many ways.

A projection can be defined by a set of premises that specify relationships between the graticule and the parallels and meridians. Some possible relationships include the following: which parallels and meridians are to be projected arc length true, at what angle will the graticule lines intersect, and which longitude and latitude will map to the graticule's origin. Premises for the Simple Cylindrical Projection could be: (i) the graticule is a rectangular grid (the projection is cylindrical), (ii) the equator and the meridians are projected arc length true (the projection is equidistant), (iii) the x-axis is the image of the equator, with eastings positive to the east (the equator is the standard parallel for the Simple Cylindrical Projection), and the y-axis is the image of the central meridian, with northings positive to the north. Choosing formulations that honor the premises on a sphere or on a reference ellipsoid creates spherical or ellipsoidal versions of a projection. Equations satisfying these premises are

$$(x,y) = f(\lambda, \phi) = (a[\lambda - \lambda_{origin}],\ a[1-\epsilon^2]\int_0^\phi [1-\epsilon^2 \sin^2 \phi]^{-3/2} d\phi) \tag{8.4}$$

To see that these equations satisfy the premises, set $\phi_\mathbf{A} = 0$ and $\phi_\mathbf{B} = \phi$ in Eq. 6.11 to give the equation for the arc length of a meridional segment s_M from the equator to geodetic latitude ϕ:

$$s_M = a(1-\epsilon^2)\int_0^\phi (1-\epsilon^2 \sin^2 \phi)^{-3/2} d\phi \tag{8.5}$$

The first premise is satisfied because $x = a(\lambda - \lambda_{origin})$ maps parallels of latitude to straight lines that are parallel to the x-axis, and $y = a(1-\epsilon^2)\int_0^\phi (1-\epsilon^2 \sin^2 \phi)^{-3/2} d\phi$ maps meridians

---

[1]**Nonmetric** means projected distances are not real distances.

8.1. DEVELOPABLE SURFACES

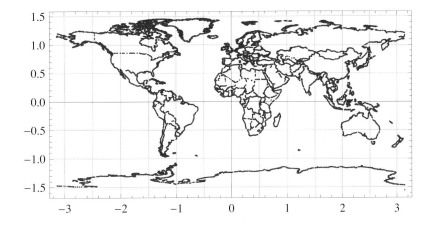

Figure 8.4: Simple cylindrical projection for a spherical Earth model with unit radius. Longitude and latitude (in radians) are treated as if they are x and y values directly. The graticule is a square grid.

to straight lines that are parallel to the y-axis. The second premise is satisfied because choosing $\phi_1 = 0$ and setting $R = a$, $x = a(\lambda - \lambda_{origin})\cos\phi_1$ maps the equator arc length true, and substituting s_M for $R\phi$ maps meridians arc length true. The third premise is satisfied because $x = a(\lambda - \lambda_{origin})$ maps the equator to the x-axis, with eastings taking their signs from $\lambda - \lambda_{origin}$, and $y = a(1 - \epsilon^2)\int_0^\phi (1 - \epsilon^2 \sin^2\phi)^{-3/2}d\phi$ maps the central meridian to the y axis, with northings taking their signs from ϕ.

Equation 8.4 produces a graticule that is not quite square because the length of a meridional quadrant is slightly less than the length of an equatorial quadrant. A different set of premises could create an ellipsoidal map projection with a square graticule, but it could not be arc length true for both the equator and the meridians.

8.1.2 Planar map projections

The *ENU* coordinate system can be used as an ellipsoidal, tangent-plane, map projection. The political boundaries of Europe, projected in an *ENU* coordinate system, are shown in Fig. 8.5.

The stereographic map projection is a planar projection that takes a line segment from, say, the North Pole, through the point of interest, and terminates it on a plane tangent to the sphere at the South Pole (see Fig. 8.6). A formula for this spherical stereographic map projection is

$$(x, y) = f(\lambda, \phi) = \left(\frac{2R\cos(\lambda - \lambda_{origin})\cos\phi}{1 - \sin\phi}, \frac{2R\sin(\lambda - \lambda_{origin})\cos\phi}{1 - \sin\phi} \right) \quad (8.6)$$

where R is the radius of the sphere, λ_{origin} is the central meridian, and (λ, ϕ) are geographic coordinates. A stereographic political boundary map is shown in Fig. 8.7. Parallels map to concentric circles, and meridians map to straight, radial lines. In Fig. 8.7, the map grid is the rectangular black grid.

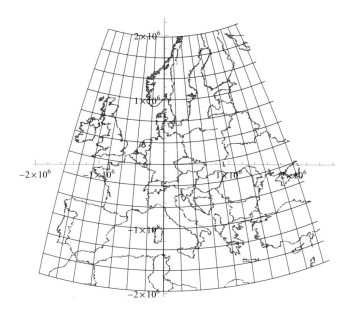

Figure 8.5: European political boundaries projected in an *ENU* coordinate system whose origin is at (10° E, 50° N).

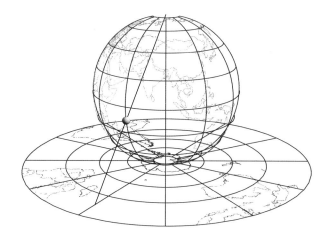

Figure 8.6: A stereographic map projection. The point of interest appears as a small ball at (10° E, 60° S).

8.1. DEVELOPABLE SURFACES

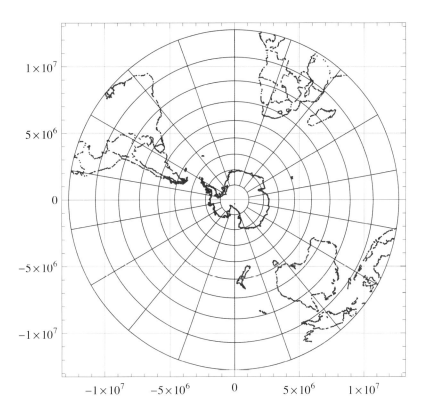

Figure 8.7: Stereographic map projection of the Southern Hemisphere.

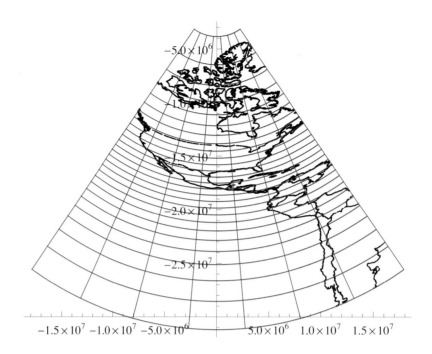

Figure 8.8: Conic map projection of a portion of the Western Hemisphere.

8.1.3 Conic map projections

Suppose a cone is placed on a sphere, whose radius is R, such that the cones's axis coincides with the sphere's axis, and the cone's apex is above the North Pole. Such a cone will touch the sphere (be tangent to the sphere) along a parallel ϕ_1. Define a map projection:

$$\theta = \lambda_{origin} - \lambda \tag{8.7}$$
$$r = -R\left[\cot \phi_1 + \cot(\pi/2 - \phi_1 + \phi)\right] \tag{8.8}$$
$$(x, y) = f(\lambda, \phi) = (r \sin \theta, r \cos \theta) \tag{8.9}$$

Equation 8.9 is the transformation from polar coordinates (r, θ) to Cartesian coordinates (x, y). In this map projection, parallels of latitudes are mapped to unequally spaced, concentric, circular arcs centered around the North Pole, with ϕ_1, and only ϕ_1, projected arc length true, and meridians are mapped to equally spaced radii. Figure 8.8 shows political boundaries between $-130° \leq \lambda \leq -60°$ and $-45° \leq \phi \leq 85°$, with $\phi_1 = 20°$.

8.2 Map Projection Types

Often a map projection is chosen for a specific application because it preserves some property without distortion. For example, navigators' maps are often compiled using Mercator's projection (Snyder 1987; Maor 1998) because curves of constant azimuth on the reference ellipsoid (loxodromes) are projected to straight lines on the map. This means that the azimuth one measures from a Mercator map using a protractor is the direction in which a craft would be steered. The importance of this property for navigation is hard to overestimate, and the maps compiled in his

8.2. MAP PROJECTION TYPES

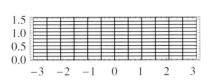
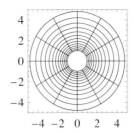

Figure 8.9: Conformal mapping (e^z) of a grid to a circle. The grid's abscissa is in the range $[-\pi, \pi]$, as for longitude expressed in radians. The ordinate is in the range $[0, \pi/2]$, the latitudes of the Northern Hemisphere.

projection made Gerardus Mercator rich and famous (Crane 2003; Taylor 2004). Over the years, scores of map projections have been created (Maling 1973; Snyder 1987; Bugayevskiy and Snyder 1995; Iliffe 2000; Qihe et al. 2000).

8.2.1 Property preserving

There are map projections that preserve one (or more) of distance, area, direction, and bearings, meaning the value of that property on the map equals the value of that property on the reference ellipsoid. Many map projections used in geodesy are **conformal**: a conformal mapping preserves bearings. The theoretical foundations of conformal map projections are rooted in complex analysis and differential geometry (Lee 1983; Lee 1974; Thomas 1979; McCleary 1994; Pressley 2007).

What does it mean to preserve bearings? Recall that bearings are the horizontal angles formed by the intersection of two curves (section 3.3.3 on page 23). Suppose these curves are the edges of a rectangular tennis court that form right-angled corners. The corners in a bearing-preserving (conformal) map projection would also be right angles. For this reason conformal projections are said to be locally shape-preserving.

Although conformal projections preserve shape locally, the overall shape might be distorted dramatically. As an example, an exceedingly simple conformal projection[2] transforms a rectangular grid into a circular one (see Fig. 8.9). Obviously, the overall shape of the rectangular grid has been dramatically changed. However, recall that conformal projections preserve shape locally by preserving angles at line intersections. The lines of the rectangular grid intersect at right angles, but so do the circles and their radii. Consider how the projected image would appear if viewed up close. The curvature of the circles would diminish until they appeared to be essentially straight-line segments, so locally the figure would still appear to be a rectangular grid. This is very common with conformal map projections. Very little discernible shape distortion occurs over small regions, but quite a lot can occur over large regions. Figure 8.10 on the next page shows a map of the Northern Hemisphere using the same projection as Fig. 8.9.

Conformal map projections have several attractive qualities:

- They preserve local shape. Conformal projections, and only conformal projections, map

---

[2] The easting and northing coordinates are the real and imaginary parts of $z = e^{\phi + i\lambda}$, respectively. That is, $(x, y) = (Re[z], Im[z])$ for $z = e^{\phi + i\lambda}$, where λ, ϕ are geographic coordinates as usual; $i = \sqrt{-1}$; and e is the base of the natural logarithm.

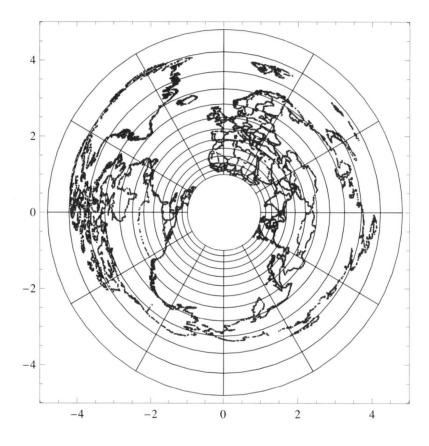

Figure 8.10: Projection of the Northern Hemisphere using the conformal mapping $z = e^{\phi + i\lambda}$. The Southern Hemisphere would be projected into the empty circle in the middle, so crushed together as to be unintelligible. The North Pole is projected to the outer blue circle, changed from a point to a curve.

8.2. MAP PROJECTION TYPES

Figure 8.11: Africa and Europe in a Lambert Conformal Conic projection (left) and in an Albers Equal-Area projection (right). Europe is very distorted in the equal-area projection and greatly enlarged in the conformal projection. This is a typical trade-off.

circles to circles everywhere on the map. Others can map circles to ellipses or even other odd shapes.

- Conformality can exist locally for other projections, even on equal-area maps. For example, the Albers Equal–Area projection is conformal along its standards (see below), but nowhere else (Snyder 1987, p. 99).

- They usually do not distort area, direction, or distance overmuch for maps of small regions.

- The mathematical analysis of conformal projections (the theory of complex variables) is comparably easier than the analysis of nonconformal projections.

- Surveyors tend to prefer conformal projections. Because nonconformal projections do not preserve shape, the horizontal angle observed by even an infinite-precision, error-free theodolite would in general not be the angle drawn on the map! This means that the scale distortion introduced by conformal projections, and only conformal projections, does not depend on azimuth (see section 8.3.3 on page 160).

All projections fail to preserve one or more, perhaps even all, of the spatial properties given above. Projections can be labeled according to which property they do not distort. Distance-preserving projections are **equidistant**, direction-preserving projections are **azimuthal**, bearing-preserving projections are **conformal**, and area-preserving projections are **equal-area** or **authalic**. Projections that preserve none of these are called **aphylactic** (Lee 1944, p. 193). No map projection can be conformal and equal-area for the whole reference ellipsoid. The Cauchy-Riemann equations guarantee that if a projection preserves shape, it distorts area and vice versa (McCleary 1994; Pressley 2007) (Fig. 8.11). Direction preservation and distance preservation do not preclude other preservation properties necessarily. Therefore, these properties do not technically create a classification for projections, although they are often described that way. Some projections called equidistant display only a subset of all distances correctly.

8.2.2 Graticule groups

Lee (1944, p. 193) writes that map projections can be classified by a "...consideration of the pattern formed by the meridians and parallels...", with the following groups: cylindric, pseudocylindric, conic, pseudoconic, and azimuthal. Lee gives these group definitions:

- Cylindric: projections in which the meridians are represented by a system of equidistant parallel straight lines, and the parallels by a system of parallel straight lines at right angles to the meridians.

- Conic: projections in which the meridians are represented by a system of equally inclined concurrent straight lines, and the parallels by concentric circular arcs, the angle between any two meridians being less than their true difference of longitude.

- Azimuthal: projections in which the meridians are represented by a system of concurrent straight lines inclined to each other at their true difference of longitude, and the parallels by a system of concentric circles with the common centre at the point of concurrency of the meridians.

These groups have often, and often incorrectly, been associated with developable surfaces (cylinders, cones, and planes). For example, Mercator's projection is in the cylindrical group, but its developable surface (which has no name) is not a cylinder. Also, the solution of the geodesic inverse problem (see section 6.1.8 on page 116) can be used as an ellipsoidal equidistant azimuthal map projection as follows. The inverse problem's solution is the geodesic distance and azimuth from a fixed point to a forward point. Distance and azimuth are the polar coordinates of the forward point with respect to the fixed point. Transforming those polar coordinates to Cartesian coordinates produces a projection of the forward point that is arc length true and at its correct azimuth. The political boundaries of Europe in this projection are shown in Fig. 8.12. This projection uses a patch of the reference ellipsoid as its developable surface, not a plane, even though it is azimuthal.

8.3 Projection Parameters

8.3.1 Aspects of projections

Projections can have direct, transverse, and oblique aspects. For the **direct aspect**, the axis of the developable surface is coincident with the minor axis of the reference ellipsoid or sphere. Figure 8.13 on the next page shows a cone in the direct aspect. For the **transverse aspect**, the axis is perpendicular to the minor axis. Figure 8.14 on page 158 shows a cylinder in the transverse aspect. For the **oblique aspect**, the axis is neither perpendicular nor parallel to the minor axis. Mathematically, cylinders and planes are special cases of cones. A cylinder is a cone whose apex has been removed to infinity, and a plane is a cone whose apex has been depressed until the figure has become planar. Therefore, all three surfaces have all three aspects. However, conic projections are not used much other than in their direct aspect, whereas cylinders and planes have common projections in all three aspects.

8.3.2 Central meridians and parallels

The geographic coordinates that get mapped to the origin of the grid $(x, y) = (0, 0)$ are called the **central meridian** and the **central parallel**. They will be denoted with a subscript as

8.3. PROJECTION PARAMETERS

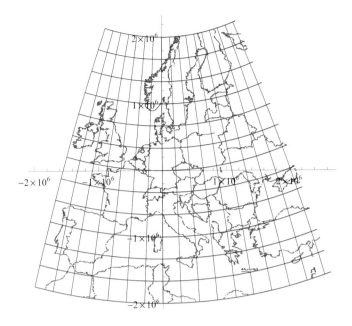

Figure 8.12: European political boundaries projected in an ellipsoidal, azimuthal, equidistant projection whose origin is at (10° E, 50° N).

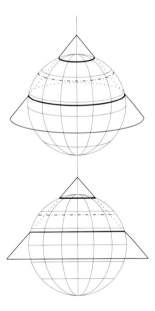

Figure 8.13: Conic projection with two standards. The standards are the heavier lines north and south of the central parallel, which is shown dashed. Note that the standards are also parallels.

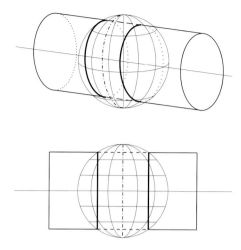

Figure 8.14: Transverse cylindrical projection with two standards. The standards are the heavier lines to the left and right of the central meridian, which is shown dashed. Note that the standards do not follow either meridians or parallels.

$(\lambda_{origin}, \phi_{origin})$, following the notation in ISO/IEC 18026(2006E). The central meridian and parallel are parameters of most map projections, and such projections can be crafted to place the central meridian and parallel as desired, sometimes in the middle of the area to be mapped and sometimes off in a corner. However, the choice of the central meridian and parallel is more a question of managing the scale distortion of the map rather than placing the grid origin in an auspicious location. That task is accomplished by using false eastings and false northings, as discussed below.

For the transverse aspect, the central meridian is that meridian whose plane is perpendicular to the axis of the developable surface (see Fig. 8.14). Only one meridian is suitable because, since the meridians are not parallel, there can be only one of them perpendicular to the axis of the developable surface. The central parallel can be chosen as desired, even at the North or South Pole. Figure 8.15 on the next page shows the South Pole mapped with the combination of two Transverse Mercator projections, one with $(\lambda_{origin}, \phi_{origin}) = (0°, -90°)$ for places with $-90° < \lambda < 90°$ and one with $(\lambda_{origin}, \phi_{origin}) = (180°, -90°)$ for places with $90° < \lambda < 270°$.

The equator is the central parallel for direct aspect cylindrical projections. Direct aspect planar projections can take either pole for the central parallel. Direct aspect conics have no obvious choice for the central parallel, and this freedom is an attractive quality of these projections, making them popular for maps of regions that are long east to west and located mostly somewhere other than the equator or the poles. For example, a conic is a natural choice for mapping Russia. The central meridian of direct aspect projections can be chosen as desired. Figure 8.16 on the facing page shows the direct aspect of a cylindrical projection, the Mercator.

Oblique aspect projections typically have neither restrictions nor obvious choices for their central parallel or meridian, a freedom that can be very attractive. Furthermore, their x- and y-axes can be oriented as desired. The y-axis is defined either by a central point and an azimuth or by two central points with the x-axis set perpendicular to the y-axis. The geographic coordinates of the grid origin are specified, but these, in general, are not the central meridian and parallel (Snyder 1987, p. 69).

8.3. PROJECTION PARAMETERS

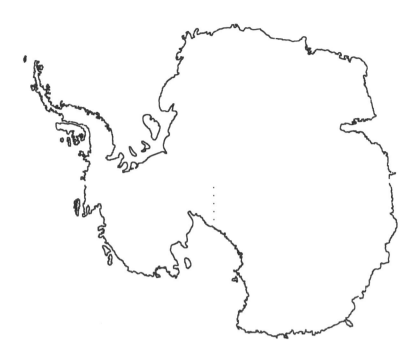

Figure 8.15: Antarctica mapped with two Transverse Mercator projections, one each for the longitude hemispheres centered around $\lambda = 0°$ and $\lambda = 180°$.

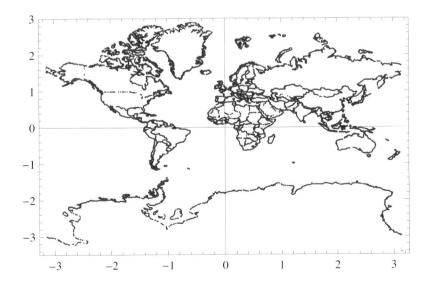

Figure 8.16: Mercator projection of the world onto a unit sphere.

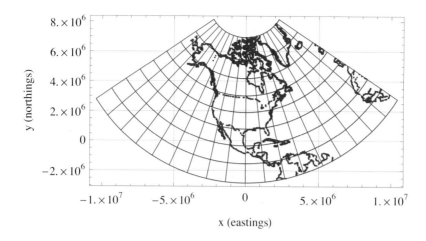

Figure 8.17: Lambert Conformal Conic projection of North America. Grid lines are black. Images of the meridians are red, and images of the parallels are blue.

8.3.3 Central scale

Eastings and northings can be multiplied by a constant k_0, called the **central scale**, which has the effect of metaphorically enlarging or reducing the map (but not the grid) on a celestial-sized photocopier. Scaling by $k_0 \neq 1$ produces a different set of curves being standard than occur when no scaling is done. Scaling by $k_0 < 1$ allows the cartographer to make scale distortions as low as possible for an area as large as possible. One approach involves crafting the projection so that $1 - k_{min} = k_{max} - 1$, meaning the minimum distortion and the maximum are equal within the area of interest. This is somewhat like having a wristwatch that is known to run 1 second fast per year and setting it to 1 second before the hour. This watch will be within ± 1 second of the true time for two years, the first year too slow and the second year too fast. This approach is appropriate for regions of small extent, but requires a special solution for large areas such as North America (Snyder 1984).

8.4 Grid Coordinates

Cartesian coordinate systems, having axes that are perpendicular straight lines, are the most common type of planimetric coordinate systems. There are other possibilities, such as polar coordinate systems, whose axes are circles and radii. However, map projection coordinate systems are always Cartesian. Map projection coordinates are referred to as **grid coordinates**.

Grid coordinates themselves are usually called "eastings" and "northings." Eastings will be denoted by x and northings by y instead of e and n for two reasons. First, this emphasizes the Cartesian nature of projected coordinates. Second, e and n already denote ENU coordinates. So why are they called eastings and northings? As shown in Fig. 8.17, cartographic grids generally do not align with the projected parallels and meridians. Nevertheless, most maps are compiled so that the x-axis is generally aligned east-west and the y-axis is generally aligned north-south. Because they usually align with geodetic directions in a general way, grid coordinates are called eastings and northings.

8.4. GRID COORDINATES

Table 8.1: Sample geographic and grid coordinates.

Point	Geographic (°)		Cylindrical (m)		Stereographic (m)	
	λ	ϕ	x	y	x	y
A	0	0	0.0	0.0	0.0	12 756 000.0
B	1	-1	111 317.1	-111 328.4	218 770.9	12 533 377.1
C	1	0	111 317.1	0.0	222 622.9	12 754 057.2

Table 8.2: Distances, azimuths, and bearings computed from geographic and grid coordinates.

Distance	Great circle (m)	Cylindrical (m)	Stereographic (m)
A,B	157 422.2	157 434.1	312 124.4
A,C	**111 317.1**	**111 317.1**	222 631.4
B,C	111 317.1	111 328.4	220 713.7
Azimuth	Spherical	Cylindrical	Stereographic
$\alpha_{A,B}$	134° 59′ 44″	135° 0′ 10″	135° 30′ 0″
$\alpha_{A,C}$	90° 0′ 0″	90° 0′ 0″	90° 30′ 0″
$\alpha_{B,C}$	0° 0′ 0″	0° 0′ 0″	1° 0′ 0″
Bearing	Spherical	Cylindrical	Stereographic
$\beta_{C,A,B}$	45° 0′ 16″	45° 0′ 10″	45° 0′ 0″
$\beta_{A,B,C}$	45° 0′ 16″	44° 59′ 50″	45° 30′ 0″
$\beta_{B,C,A}$	90° 0′ 0″	90° 0′ 0″	89° 30′ 0″

Maps are used to determine distance, area, direction, and bearing. These will be illustrated in turn by comparing some results computed on a sphere with $R = 6378$ km and those computed from the grid coordinates. The following points will be used: **A** = (0° E, 0° N), **B** = (1° E, 1° S), and **C** = (1° E, 0° N). The grid coordinates in Table 8.1 come from the cylindrical projection (Eq. 8.1) and the stereographic projection (Eq. 8.6).

Grid coordinates are Cartesian, so grid distances d_g are computed using Pythagoras's formula, Eq. 3.6. Distances computed with geographic coordinates on a sphere are great circle distances (Eq. 6.6). Table 8.2 gives distances between the point pairs. Great circle distances should be taken as the correct values because they are geodesic distances for a sphere. The grid distance in boldface matches the correct value. The only correct grid distance is the equatorial segment in the cylindrical projection, which projects the equator arc length true (see section 8.1.1 on page 145). The distances computed with the cylindrical and stereographic coordinates are very different from each other. This example illustrates that, in general, grid distances are not the same as geodesic distances but sometimes they can be.

Grid azimuths α_g come from the arctangent of the ratio of the coordinate differences, Eq. 3.8. Azimuths on the sphere from **X** to **Y** are denoted $\alpha_{X,Y}$. Two of the azimuths between **A**, **B**, and **C** can be deduced by inspection. $\alpha_{A,C} = 90°$ because the points have the same latitude and **C** is east of **A**. $\alpha_{B,C} = 0°$ because the points have the same longitude and **C** is north of **B**. $\alpha_{A,B}$ cannot be deduced by inspection, so we'll use the spherical azimuth formula (Eq. 7.20):

$$\sin \alpha_{A,B} = \frac{\sin \theta_B \sin \delta\lambda}{\sin c}$$

In this equation $\theta_\mathbf{B}$ is the spherical colatitude of **B**, so $\theta_\mathbf{B} = 90° - \phi_\mathbf{B} = 91°$. c is the great circle distance between the points, which was computed to be $157\,422.2$ m. The azimuth equation was developed on a unit sphere, so the great circle distance must be divided by the sphere's radius to provide the equivalent distance on a unit sphere: $c/R = 157\,422.2/6\,378\,000 = 0.024\,6821$. $\delta\lambda = \lambda_\mathbf{B} - \lambda_\mathbf{A} = 1° - 0° = 1°$.

$$\begin{aligned}\sin\alpha_{\mathbf{A},\mathbf{B}} &= \frac{\sin\theta_\mathbf{B}\,\sin\delta\lambda}{\sin c} \\ &= \frac{\sin(91°)\,\sin(1°)}{\sin(0.024\,6821)} \\ &= -0.707\,053 \\ \alpha_{\mathbf{A},\mathbf{B}} &= \arcsin(-0.707\,053) \\ &= -44°\,59'\,44''\end{aligned}$$

Disambiguating for quadrant gives $\alpha_{\mathbf{A},\mathbf{B}} = 135°\,0'\,16''$.

Table 8.2 also gives the azimuths between the point pairs. As before, the spherical azimuths should be taken as the correct values. The cylindrical azimuths are correct along meridians and parallels. No other grid azimuth is correct.

Bearings are the differences of azimuths. For example, $\beta_{\mathbf{CAB}} = \alpha_{\mathbf{AB}} - \alpha_{\mathbf{AC}}$. Only one grid bearing in Table 8.2 is correct. The grid bearings sum to 180°, as they must according to the laws of plane geometry. The spherical bearings sum to $180°\,0'\,32''$. The sum of a spherical triangle's interior angles always exceeds 180°. The excess beyond 180° is called **spherical excess**, so the spherical excess here is $32''$.

The mathematical formulation of projections takes $(\lambda_{origin}, \phi_{origin}) \to (x,y) = (0,0)$. Allowing the coordinate values of the grid origin to remain (0,0) generally results in undesirable negative grid coordinates. To avoid this, the coordinates of the grid origin can be assigned arbitrary, convenient values – **false easting** and **false northing** – to ensure that no location in the area of interest has a negative grid coordinate value. False eastings and northings are simply added to the coordinates that are computed from the projection formulæ.

8.4.1 Transverse Mercator

The following premises define the Transverse Mercator (TM) projection: (i) the mapping is conformal, (ii) the central meridian is projected arc length true in all its parts, (iii) the central meridian maps to the y-axis, positive to the north, and (iv) the grid origin corresponds to the point of intersection of the central meridian and the equator. TM projections are often used where the extent of a region is longer north to south than east to west because, since the central meridian is projected arc length true, the projection suffers less scale distortion close to the central meridian.

To project geodetic longitude $-\pi \le \lambda < \pi$ to TM easting x and geodetic latitude $-\pi/2 < \phi <$

8.4. GRID COORDINATES

$\pi/2$ to TM northing y (Snyder 1987, p. 61)

$$x_{TM} = N[A + (1 - T + C)A^3/6 + (5 - 18T + T^2 + 72C - 58(\epsilon')^2)A^5/120] \quad (8.10a)$$

$$y_{TM} = \{d - d_{origin} + N \tan\phi[A^2/2 + (5 - T + 9C + 4C^2)A^4/24 + \\ (61 - 58T + T^2 + 600C - 330(\epsilon')^2)A^6/720]\} \quad (8.10b)$$

$$k_{TM} = [1 + (1 + C)A^2/2 + (5 - 4T + 42C + 13C^2 - 28(\epsilon')^2)A^4/24 + \\ (61 - 148T + 16T^2)A^6/720] \quad (8.10c)$$

$$T = \tan^2\phi \quad (8.10d)$$

$$C = (\epsilon')^2 \cos^2\phi \quad (8.10e)$$

$$A = (\lambda - \lambda_{origin})\cos\phi \quad (8.10f)$$

$$d = a\,[(1 - \epsilon^2/4 - 3\epsilon^2/64 - 5\epsilon^6/256 - \ldots)\phi - (3\epsilon^2/8 + 3\epsilon^4/32 + 45\epsilon^6/1024 + \ldots)\sin 2\phi + \\ (15\epsilon^4/256 + 45\epsilon^6/1024 + \ldots)\sin 4\phi - (35\epsilon^6/3072 + \ldots)\sin 6\phi + \ldots] \quad (8.10g)$$

where a is the length of the semimajor axis, ϵ is eccentricity, ϵ' is second eccentricity, N is radius of curvature in the prime vertical at ϕ, λ_{origin} is the central meridian, and ϕ_{origin} is the central parallel. d is a truncated infinite series expansion of Eq. 8.5 (meridional arc length from the equator to ϕ). d_{origin} is d at ϕ_{origin}. The error in d grows essentially linearly from zero at the equator to about 3000 m too large at the poles, a maximum error of 0.05% for the WGS 84 reference ellipsoid. Even so, this formulation of the TM projection loses numerical accuracy very rapidly for longitudes outside the range $\lambda_{origin} \pm 3°$. Modern approaches do not suffer from these longitude restrictions (Bermejo-Solera and Otero 2009).

A false easting $e_\mathcal{F}$, a false northing $n_\mathcal{F}$, and a central scale factor k_0 modify x_{TM}, y_{TM}, and k_{TM} as follows:

$$x = k_0\,x_{TM} + e_\mathcal{F} \quad (8.11a)$$

$$y = k_0\,y_{TM} + n_\mathcal{F} \quad (8.11b)$$

$$k = k_0\,k_{TM} \quad (8.11c)$$

8.4.2 Lambert Conformal Conic

Regions that extend longer east-to-west than north-to-south are often mapped with conic projections because their standards run east to west. This creates a low scale-distortion strip around the central parallel; a naturally east-west region (see Fig. 8.13 on page 157).

Lambert Conformal Conic (LCC) map projections are widely used in mapping and geodesy. They can have one or two standard parallels. When there are two standard parallels, they are denoted ϕ_1 and ϕ_2. To project geodetic longitude $-\pi \leq \lambda < \pi$ to LCC easting x and geodetic latitude $-\pi/2 < \phi < \pi/2$ to LCC northing y (Snyder 1987, p. 107)

$$x_{LCC} = \rho \sin\theta \quad (8.12a)$$

$$y_{LCC} = \rho_{origin} - \rho \cos\theta \quad (8.12b)$$

$$k_{LCC} = \rho\,n/(am) \quad (8.12c)$$

where

$$\rho = aFt^n \tag{8.12d}$$
$$\theta = n(\lambda - \lambda_{origin}) \tag{8.12e}$$
$$n = (\ln m_1 - \ln m_2)/(\ln t_1 - \ln t_2) \tag{8.12f}$$
$$m = \cos\phi(1 - \epsilon^2 \sin^2 \phi)^{1/2} \tag{8.12g}$$
$$t = \tan(\pi/4 - \phi/2)/[(1 - \epsilon\sin\phi)/(1 + \epsilon\sin\phi)]^{\epsilon/2} \tag{8.12h}$$

or

$$= \left[\left(\frac{1-\sin\phi}{1+\sin\phi}\right)\left(\frac{1+\epsilon\sin\phi}{1-\epsilon\sin\phi}\right)^{\epsilon}\right]^{1/2} \tag{8.12h'}$$
$$F = m_1/(nt_1^n) \tag{8.12i}$$

and a is the length of the semimajor axis, ϵ is eccentricity, $t_x = t$ at ϕ_x, $m_x = m$ at ϕ_x, and $\rho_{origin} = aFt_{origin}^n$. A false easting $e_{\mathcal{F}}$ and a false northing $n_{\mathcal{F}}$ modify x_{LCC} and y_{LCC} as follows:

$$x = x_{LCC} + e_{\mathcal{F}} \tag{8.13a}$$
$$y = y_{LCC} + n_{\mathcal{F}} \tag{8.13b}$$

Convergence γ_{LCC} is computed by

$$\gamma_{LCC} = (\lambda - \lambda_{origin})\sin\phi_0 \tag{8.14}$$

where

$$\sin\phi_0 = \frac{\ln[W_2\cos\phi_1/(W_1\cos\phi_2)]}{Q_2 - Q_1}$$
$$W = (1 - \epsilon^2\sin^2\phi)^{1/2}$$
$$Q = \frac{1}{2}\left[\ln\frac{1+\sin\phi_1}{1-\sin\phi_1} - e\ln\frac{1+e\sin\phi_1}{1-e\sin\phi_1}\right]$$

e is the base of the natural logarithm, and $W_1 = W$ at ϕ_1, and so on.

8.5 Map Projection Systems

Producing maps of a common area according to a system of rules that guarantee that those maps form a consistent representation of the features in that area can provide enormous benefits. Various mapping organizations have created standardized mapping systems, with scales of applicability ranging from global to local. A **map projection system** consists of conventions that stipulate how geodetic coordinates are transformed to and from grid coordinates by way of map projections. Some systems use only one kind of projection; others use several. Systems typically have an allowable scale distortion maximum or minimum, which dictate how far from the standards the system should be used. Keeping the maximum scale distortion acceptably low requires large areas be divided into **zones**. Map projection systems usually have rules for labeling zones, formating eastings and northings, applying false eastings and northings, and using units of measure.

8.5.1 Universal Transverse Mercator (UTM)

The Universal Transverse Mercator map projection system is a product of the U.S. Department of Defense (Hager et al. 1989). UTM divides the Earth into 60 zones by longitude, each spanning 6° west to east, with some small exceptions. Each longitudinal band is split south to north at the equator, for a total of $60 \times 2 = 120$ zones.

Zones are labeled 1 - 60 according to their longitude and either an N or an S for hemisphere. Zone 1 spans from 180° W to 174° W (180° E to 186° E). Zone numbers increase sequentially to the east, so zone 2 covers from 174° W to 168° W (186° E to 192° E). The lowest-number zone in the Eastern Hemisphere is zone 30, which covers from 0° E to 6° E, and zone 31 covers from 6° E to 12° E. The highest-number zone is zone 60, spanning from 174° E to 180° E.

Southern Hemisphere zones span from 80° S to the equator. Northern zones span from the equator to 84° N. The Universal Polar Stereographic (UPS) map projection system can be used to map polar regions (Hager et al. 1989).

Geodetic coordinates λ, ϕ produce UTM eastings and northings x, y by ellipsoidal Transverse Mercator projections, but no particular reference ellipsoid is stipulated. The Transverse Mercator projections are parameterized to fit each zone individually. The central meridians λ_{origin} of the projections are the central meridians of the zones, being 3° of longitude east of the zones' western boundaries (and 3° of longitude west of the zones' eastern boundaries). For all zones, the equator is the central parallel ($\phi_{origin} = 0°$), the central scale factor is $k_0 = 0.9996$, and the false easting is $e_\mathcal{F} = 500\,000$ m. For the WGS 84 reference ellipsoid, $a = 6\,378\,137$ m, so zones extend $6\,378\,137 \cdot 3° \cdot \pi/180° = 333\,958$ m east and west of the central meridian at the equator. Eastings decrease to the west, so the smallest WGS 84 UTM easting is $500\,000 - 333\,958 = 166\,042$ m. Choosing 500 000 guarantees that eastings within the zone are not negative.

Northern Hemisphere zones have a false northing of 0 m, whereas Southern Hemisphere zones have a false northing of $10\,000\,000$ m. Since $k_0 = 0.9996$, UTM does not preserve arc length along the central meridian. Thus, a 1-m change along the central meridian produces a 0.9996 m change in UTM northings. Northings increase northwards in all zones. Since $10\,000\,000$ m is longer than the length of the meridional arc from 0° to 80° S, northings cannot be negative in Southern Hemisphere zones.

UTM coordinates are computed by first determining in which zone the point belongs, which establishes the projection parameters. Then $\lambda, \phi \to x, y$ by the Transverse Mercator projection (see section 8.4.1). UTM coordinates are always given in meters and include their zone, e.g., 18N 5 331 201, 4 656 239. On an NGS data sheet, the UTM coordinates computed on GRS 80 look like this (the dash to the left of the northing is a space separator, not a minus sign):

```
LX3030;                    North          East      Units Scale Factor Converg.
LX3030;UTM  18      -  4,632,606.701   728,378.794   MT    1.00024186   +1 50 01.8
```

Important UTM features can be summarized as follows:

- Zone 1 and zone 60 straddle the 180° meridian, not the prime meridian. Zone numbers increase and decrease sequentially except at the 180° meridian.

- The boundary between zone 60 and zone 1 does not follow the 180° meridian exactly. Instead, it zigzags across it to accommodate international boundaries. Also, certain zone boundaries in northern Europe, such as near Iceland, have been adjusted so that important land masses are in a single zone.

- For all zones, places on the equator have two different northing coordinates: either 0 for the northern zones or 10 000 000 for the southern zones.

- UTM coordinates are large numbers. Eastings are always on the order of hundreds of thousands and northings in the middle latitudes (where human populations are concentrated) are always in the millions.

- Geodetic (ellipsoid) height is not necessarily carried along as a UTM height coordinate. Grid coordinates often do not have height at all, and if they do, the height is likely to be orthometric (see section 10.1.1 on page 191).

- UTM is usually not suitable for land surveying because the most extreme projection scale factors can introduce linear distortions that exceed typical land surveying tolerances.

8.5.2 State Plane Coordinate System (SPCS)

States individually adopt the U.S. State Plane Coordinate System (Stem 1995) by statute under the guidance of and in cooperation with the U.S. National Geodetic Survey. Each state is divided into one or more zones whose boundaries follow administrative boundaries. Each zone is assigned a projection based on whether it is long east to west (Lambert Conformal Conic), north to south (Transverse Mercator), or some other direction (Oblique Mercator). Some states have zones running in two directions (e.g., Florida). Alaska has the most zones, including the only Oblique Mercator zone for its southern panhandle. Projection parameters are usually chosen to ensure that scale distortion is less than 1:10 000 in order to make SPCS zones suitable for most land surveying tasks. Zones are labeled with the state name and an adjective indicating location, such as Arizona Central. SPCS83 zones also are identified with unique numeric codes, such as 0600 for the Connecticut zone.

The SPCS was first implemented under NAD 27, and was later reimplemented for NAD 83. The NAD 83 implementation usually left the zone definitions the same as for NAD 27, but some states changed how many zones they have and, in one case, even which direction they run. Each zone has its own origin with assigned false easting and northing coordinates. All states adopted U.S. survey feet in NAD 27 but switched to meters in NAD 83. The values of the NAD 83 grid origin in meters are, in some cases, a conversion from the NAD 27 value in U.S. Survey Feet resulting in numbers such as 304 800.6096 m for the Connecticut zone false easting. The exact definition of the SPCS is extensive, involving hundreds of projection parameter definitions (Stem 1995).

Example 8.1. A total station is set up at station KERR at 72° 14′ 54″.344 03 W, 41° 48′ 42″.473 64 N, 152.503 m. After reductions from natural to geometric observations, station ALDRICH is observed to be at a mark-to-mark slope distance $d_s = 148.531$ m, a mark-to-mark zenith angle of $\zeta_{K,A} = 88° 47′ 50″$, and an azimuth of $\alpha_A = 329° 25′ 20″$. Use coordinate transformations to determine ALDRICH's SPCS83 (0600) coordinates.

Solution: This solution strategy depends on these coordinate system transformations and map projections: $ENU \to XYZ \to LBH \to$ SPCS83.

1. $ENU \to XYZ$: The local origin is at KERR. Using plane surveying at the local origin, the ENU coordinates of ALDRICH are (-75.541, 127.848, 3.117). From Eq. 5.5, the XYZ coordinates of ALDRICH are (1 451 508.754, -4 534 331.620, 4 230 245.348) in meters.

2. $XYZ \to LBH$: From Eq. 5.2, the LBH coordinates of ALDRICH are 72° 14′ 57″.616 78 W, 41° 48′ 46″.617 36 N, 155.622 m.

8.5. MAP PROJECTION SYSTEMS

3. $LBH \to$ SPCS83: SPCS83 zone 0600 uses a Lambert Conformal Conic projection, the GRS 80 reference ellipsoid, central meridian $\lambda_{origin} = 72° 45'$, central parallel $\phi_{origin} = 40° 50'$, standard parallels at $\phi_1 = 41° 12'$ and $\phi_2 = 41° 52'$, false easting $e_\mathcal{F} = 304\,800.6096$ m, and false northing $n_\mathcal{F} = 152\,400.3048$ m (Stem 1995). From Eq. 8.12, the SPCS(0600) coordinates of ALDRICH are (346 401.654, 261 317.340). The ellipsoid height can be associated with this position in order to be able to reverse all the transformations.

□

Example 8.2. With the same data in Example 8.1, use reductions to determine ALDRICH's SPCS83 (0600) coordinates.

Solution:

1. Determine SPCS83 grid distance d_g. No orthometric heights were provided, so we proceed with the ellipsoid-height formulæ.

    (a) Estimate $h_\mathbf{A}$ with a vertical distance:
    $$h_\mathbf{A} \approx h_\mathbf{K} + d_s \sin \zeta_{\mathbf{K},\mathbf{A}}$$
    $$= 152.503 + 148.531 \sin 88° 47' 50''$$
    $$= 155.621$$

    (b) Reduce d_s to a normal section chord d_c using Eq. 6.15, and then reduce d_c to a normal section arc d_{ns} using Eq. 6.16. These equations depend on the average radius of curvature in the normal section $\overline{R}_\alpha$ (Eq. 4.9), which depends on the meridional radius of curvature M and the prime vertical radius of curvature N. Using Eq. 4.7, $M = 6\,363\,821.057$ m. Using Eq. 4.8, $N = 6\,387\,647.149$ m. Then
    $$\frac{1}{R_\alpha} = \frac{\cos^2 \alpha}{M} + \frac{\sin^2 \alpha}{N}$$
    $$= \frac{\cos^2 329° 25' 20''}{6\,363\,821.057} + \frac{\sin^2 329° 25' 20''}{6\,387\,647.149}$$
    $$R_\alpha = 6\,369\,969.790$$

    KERR and ALDRICH are not very far apart, so $\overline{R}_\alpha \approx R_\alpha$.
    $$d_c = \sqrt{\frac{d_s^2 - (h_\mathbf{A} - h_\mathbf{K})^2}{(1 + \frac{h_\mathbf{K}}{R_\alpha})(1 + \frac{h_\mathbf{A}}{R_\alpha})}}$$
    $$= \sqrt{\frac{148.531^2 - (155.621 - 152.503)^2}{(1 + \frac{152.503}{6\,369\,969.790})(1 + \frac{155.621}{6\,369\,969.790})}}$$
    $$= 148.495$$

    The length of the normal section arc d_{ns} is
    $$d_{ns} = 2\overline{R}_\alpha \arcsin(\frac{d_c}{2\overline{R}_\alpha})$$
    $$= 2 \times 6\,369\,969.790 \arcsin(\frac{148.495}{2 \times 6\,369\,969.790})$$
    $$= 148.495$$

(c) Use Eq. 8.12c to compute LCC scale factor k at KERR, or obtain k from a coordinate calculator such as NADCON: $k = 0.999\,994\,899$.

(d) Reduce d_{ns} to d_g by $d_g = k\,d_{ns} = 0.999\,994\,899 \times 148.495 = 148.494$ m.

2. Determine SPCS83 grid azimuth α_g.

   (a) Compute LCC convergence $\gamma_\mathbf{K}$ using Eq. 8.14. For zone 0600, $\lambda_{origin} = 72°\,45'$ W, and $\phi_{origin} = 40°\,50'$, which leads to $\gamma_\mathbf{K} = 0°\,19'\,57\rlap{.}''257\,27$.

   (b) Solving for α_g in Eq. 7.21:
   $$\alpha_g = \alpha - \gamma_\mathbf{K} = 329°\,25'\,20'' - 0°\,19'\,57\rlap{.}''257\,27 = 329°\,5'\,22\rlap{.}''742\,732.$$

3. Determine the SPCS83 easting and northing for ALDRICH. KERR's SPCS83 coordinates are obtained from a coordinate calculator or by projecting using Eq. 8.13. They are $(e_\mathbf{K}, n_\mathbf{K}) = (346\,477.934, 261\,189.935)$ in meters.

   (a) Compute $e_\mathbf{A}$ from Eq. 3.17a:
   $$\begin{aligned} e_\mathbf{A} &= e_\mathbf{K} + d_g \sin(\alpha_g) \\ &= 346\,477.934 + 148.495 \sin(329°\,5'\,22\rlap{.}''742\,732) \\ &= 346\,401.653 \end{aligned}$$

   (b) Compute $n_\mathbf{A}$ from Eq. 3.17b:
   $$\begin{aligned} n_\mathbf{A} &= n_\mathbf{K} + d_g \cos(\alpha_g) \\ &= 261\,189.935 + 148.495 \cos(329°\,5'\,22\rlap{.}''742\,732) \\ &= 261\,317.339 \end{aligned}$$

The coordinates computed by reductions differ from the coordinates computed by transformations by 1 mm each.

□

8.6 Your Turn

Problem 8.1. A total station is set up at station KERR at $72°\,14'\,54\rlap{.}''344\,03$ W, $41°\,48'\,42\rlap{.}''473\,64$ N, 152.503 m. After reductions from natural to geometric observations, station PALMER is observed to be at a mark-to-mark slope distance $d_s = 100.100$ m, a mark-to-mark zenith angle of $\zeta_{\mathbf{K},\mathbf{P}} = 92°\,34'\,16''$, and an azimuth of $\alpha_\mathbf{P} = 204°\,57'\,50''$. Use coordinate transformations to determine PALMER's SPCS83 (0600) coordinates.

Problem 8.2. Determine PALMER's SPCS83 0600 coordinates from the data in problem 8.1 using reductions this time instead of coordinate system transformations. The SPCS83 projection scale factor $k = 0.999\,994\,8668$ at KERR, and the convergence $\gamma = -0°\,19'\,57\rlap{.}''26$.

Part III

Physical Geodesy

Chapter 9

Gravity, Geopotential, and the Geoid

Most people think of height as meaning a distance from sea level. Sea level's location comes from the Earth's gravity field, so we study gravity in order to understand height. Gravity and height systems are the concern of physical geodesy. The Geodetic Glossary (NGS 2009) defines **height** as the "distance, measured along a perpendicular, between a point and a reference surface, e.g., the height of an airplane above the ground." Although this definition seems to capture the intuition behind height very well, it contains a (deliberate) ambiguity regarding the reference surface from which the measurement is made.

Heights fall broadly into two categories: those defined with reference to the Earth's gravity field, and those defined using geometry alone. These heights are not directly interchangeable. They not only have different reference surfaces, they mean different things. As will be explained in chapter 10, in the absence of site-specific gravity measurements, no rigorous transformation between them can be accomplished.

The Geodetic Glossary (NGS 2009) contains 17 definitions for "elevation" and 23 definitions for "height," including elevations, orthometric heights, dynamic heights, geopotential numbers, normal heights, and ellipsoid heights. The definitions and relationships among these heights can be confusing, but they can appear on an NGS data sheet, so geomatics practitioners need to become familiar with them.

9.1 Mean Sea Level

Orthometric elevations and orthometric heights are probably the most widely known types of height. The Geodetic Glossary (NGS 2009) refers **height, orthometric** to **elevation, orthometric**, which is defined as the "distance between the geoid and a point measured along the plumb line and taken positive upward from the geoid." Notice that the distance is measured from the geoid and not from mean sea level. For contrast, here is the first definition for **elevation**:

> The distance of a point above a specified surface of constant potential; the distance is measured along the direction of gravity between the point and the surface.
>
> The surface usually specified is the geoid or an approximation thereto. Mean sea level was long considered a satisfactory approximation to the geoid and therefore suitable for use as a reference surface. It is now known that mean sea level can differ from the geoid

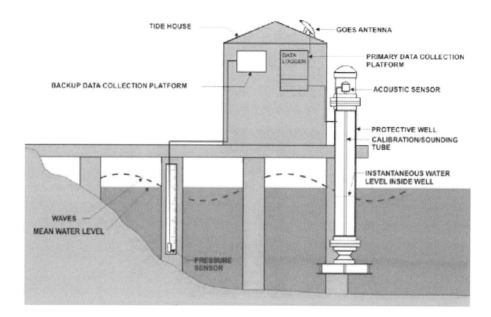

Figure 9.1: NOAA tide house and tide gauge (courtesy of NOAA 2007).

by up to a meter or more, but the exact difference is difficult to determine.

The terms *height* and *level* are frequently used as synonyms for elevation. In geodesy, height also refers to the distance above an ellipsoid...

Lying within these two definitions is a remarkably complex situation involving the Earth's gravity field and our attempts to make measurements using it as a frame of reference.

Using mean sea level for a height reference surface is perfectly natural because most human activity occurs at or above sea level. However, creating a workable and repeatable mean sea level reference surface is somewhat subtle. The Geodetic Glossary (NGS 2009) defines mean sea level to be the "average location of the interface between ocean and atmosphere, over a period of time sufficiently long so that all random and periodic variations of short duration average to zero." The National Oceanic and Atmospheric Administration's (NOAA) National Ocean Service (NOS) Center for Operational Oceanographic Products and Services (CO-OPS) has set 19 years as the period suitable for measurement of mean sea level at tide gauges (NGS 2009). Hicks (2006, p. 52) writes, "A series length of 19 years was chosen because it contains all the tidal periods that need to be considered in tidal computations; i.e., through the 18.61-year node cycle..." (see Doodson 1922; Bomford 1962; Bearman 1999; Zilkoski 2001; Boon 2004; Hicks 2006 for more details about mean sea level and tides).

Local mean sea level is often measured using a tide gauge, which is an instrument that measures the instantaneous sea level. Figure 9.1 depicts a tide house, which contains equipment that collects and records readings from a tide gauge.

It has been suspected at least since the time of the building of the Panama Canal that mean sea level might not be at the same height everywhere (McCullough 1978). The original canal, attempted by the French, was to be cut at sea level, and there was concern that the Pacific Ocean

9.1. MEAN SEA LEVEL

might not be at the same height as the Atlantic, thereby causing a massive flood through the cut. This concern became irrelevant when the sea level approach was abandoned. However, the subject surfaced again in the creation of the National Geodetic Vertical Datum of 1929 (NGVD 29) (Berry 1976, p. 143).

By this time it was known that not all mean sea level stations were the same height. This might seem counterintuitive because mean sea level stations are at an elevation of zero by definition, water seeks its own level, and the oceans have no visible constraints preventing free flow between the stations (apart from the continents). The answer lies in differences in temperature, chemistry, ocean currents, and ocean eddies.

The water in the oceans is constantly moving at all depths. Seawater at different temperatures contains different amounts of salt and, consequently, has density gradients. These density gradients give rise to immense deep-ocean cataracts that constantly transport massive quantities of water from the poles to the tropics and back (Broecker 1983; Whitehead 1989; Ingle 2000). The Sun's warming of sea surface waters creates global-scale currents that are well known to mariners in addition to other more subtle effects (Chelton et al. 2004). Geostrophic effects cause large-scale, persistent ocean eddies that push water against or away from the continents, depending on the direction of the eddy's circulation. These effects can create sea surface topographic variations of more than 50 cm (JPL 2004). As described by Zilkoski (2001, p. 40), the differences are due to "... currents, prevailing winds and barometric pressures, water temperature and salinity differentials, topographic configuration of the bottom in the area of the gauge site, and other physical causes ..."

In essence, these factors push the ocean's surface and hold it upshore, or away from the shore, further than would be the case under the influence of gravity alone. The persistent nature of these climatic factors prevents the elimination of their effects by averaging (Speed, Jr et al. 1996b; Speed, Jr. et al. 1996a). This gives rise to the seemingly paradoxical situation that holding one sea level station as a zero height reference and running levels to another station generally indicates that the other sea level station is not also at zero height, even in the absence of experimental error. Similarly, measuring the height of an inland bench mark using two level lines that start from different tide gauges generally results in two statistically different height measurements.

For example, since salinity in the oceans varies from place to place, the density of the water in the oceans is not constant, either, because it depends on the salinity. Suppose there are static water columns along a coastline, and there is constant gravitational acceleration. The water in column A in Fig. 9.2 is less dense than in column B; perhaps a river empties into the ocean at that place. The water columns must have the same weight, or else water will flow from the heavier column to the lighter column. It takes more water of lesser density to have the same mass as a column of denser water. Water is nearly incompressible, so water column A must be taller than water column B. Therefore, a mean sea level station at A would not be at the same distance from the Earth's center of gravity as a mean sea level station at B. The difference between the sea surface and the geoid is called **dynamic topography**.

Actually, the geoid is the most natural reference surface for heighting. A comprehensive understanding of the geoid requires a serious inquiry (Blakely 1995; Bomford 1962; Heiskanen and Moritz 1967; Kellogg 1953; Ramsey 1981; Torge 1997; Vaníček and Krakiwsky 1996), but the underlying concepts can be grasped without examining all the details. The heart of the matter lies in the relationship between gravitational force and gravitational potential. To understand this relationship, we need to review the concepts of force, work, and energy.

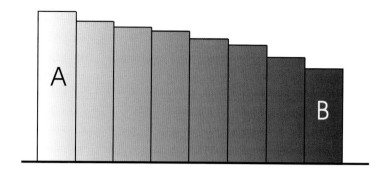

Figure 9.2: A collection of water columns whose salinity, and therefore density, has a gradient from left to right.

9.2 Physics

Gravity and gravitation are fundamental concepts in physical geodesy. **Gravitation** is the force created by mass. **Gravity** includes gravitational force but also includes acceleration from the Earth's rotation.

9.2.1 Gravity and gravitation

Force is what makes things go. This is apparent from Newton's law, $\mathbf{F} = m\mathbf{a}$, which says that the acceleration of an object is caused by, and is in the direction of, a force $\mathbf{F}$ and is inversely proportional to the object's mass m. Force has magnitude (i.e., strength) and direction, so it is represented mathematically as a vector. We denote vectors in boldface, either upper- or lowercase (e.g., $\mathbf{F}$ or $\mathbf{f}$), and scalars in lightface italics (e.g., the speed of light is commonly denoted as c). Force is measured in newtons (N), which are mass times length per second squared.

Newton's law of gravitation specifies that the gravitational force exerted by a point mass M on another point mass m is

$$\mathbf{F}_g = -\frac{GMm\hat{\mathbf{r}}}{|\mathbf{r}|^2} \tag{9.1}$$

where G is the universal gravitational constant, $\mathbf{r}$ is a vector from M's center of mass to m's center of mass, and $\hat{\mathbf{r}}$ is a unit vector in the direction of $\mathbf{r}$. The negative sign accounts for gravitation being an attractive force by orienting $\mathbf{F}_g$ in the direction opposite of $\hat{\mathbf{r}}$. Because $\hat{\mathbf{r}}$ is directed from M to m, the negative sign means $\mathbf{F}_g$ is directed from m to M. Equation 9.1 indicates that the magnitude of gravitational force is in proportion to the masses of the two objects, inversely proportional to the square of the distance separating them, and that the force is directed along the straight line joining their centroids.

In physical geodesy, M (not to be confused with the radius of curvature in the meridian) usually denotes the mass of the Earth (including the oceans and atmosphere), and consequently, the product GM arises frequently. Although the values for G and M are known independently (G has a value of approximately 6.67259×10^{-11} m^3 s^{-2} kg^{-1} and M is approximately 5.9737×10^{24} kg), their product can be measured as a single quantity, and its value has been determined to have several, nearly identical values, such as $GM = (398\,600\,441.5 \pm 0.8) \times 10^6$ m^3 s^2 (Groten 2004).

9.2. PHYSICS

Gravitational force forms a force field, meaning that the gravitational force created by any mass permeates all of space. One consequence of the principle of superposition is that gravitational fields created by different masses are independent of one another (Schey 1992). Therefore, the gravitational field created by a single mass can be examined without taking into consideration any objects within that field. Equation 9.1 can be modified to describe a gravitational field simply by dividing both sides of the equation by a unit test mass $m = 1$. We can compute the strength of the Earth's gravitational field at a distance equal to the equatorial radius (6 378 137 m) from the center of M by

$$\mathbf{E}_g = -\frac{GM\hat{\mathbf{r}}}{|\mathbf{r}|^2} \qquad (9.2)$$

$$= -\frac{398\,600\,441.5 \text{ m}^3/\text{s}^2 \hat{\mathbf{r}}}{(6\,378\,137 \text{ m})^2}$$

$$= 9.798\,29 \text{ m/s}^2(-\hat{\mathbf{r}}) \qquad (9.3)$$

This value is slightly larger than the well known value of 9.780 33 m/s^2 because the latter includes the effect of the Earth's rotation. The gravity experienced on and in the Earth is a combination of the gravitation produced by its mass and the centrifugal force created by its rotation. Notice that Eq. 9.2 has units of acceleration, not force, by virtue of division by m.

Equation 9.2 can be used to draw a picture that captures, to some degree, the shape of the Earth's gravitational field. The vectors in Fig. 9.3 on the next page indicate the magnitude and direction of force that would be experienced by a unit mass located at that point in space. The vectors decrease in length as distance from the Earth increases, and are directed radially toward the Earth's center, as expected. However, the Earth's gravitational field pervades all of space; it is not discrete as the figure suggests. Furthermore, in general, any two points in space experience different gravitational forces, if only in direction.

9.2.2 Normal gravity

The Earth's rotation causes centrifugal force that acts to accelerate objects on the Earth in a direction perpendicularly away from the axis of rotation. This acceleration is indistinguishable from gravitational acceleration, and the two combined create Earth's gravity field. Therefore, gravity depends on latitude but not longitude if the Earth is taken to be a body of homogeneous mass distribution. Under these simplifying assumptions, Newton's law can be used to create a model of gravity acceleration, called **normal gravity**. Somiglinana (1929) developed the first rigorous formula for normal gravity (also, see Heiskanen and Moritz 1967, p. 70, Eq. 2-78 and NIMA 1997, p. 4-1, Eq. 4-1):

$$\gamma = \gamma_e \frac{1 + k \sin^2 \phi}{\sqrt{1 - \epsilon^2 \sin^2 \phi}} \qquad (9.4)$$

where $k = b\gamma_p/(a\gamma_e) - 1$, a and b are the semimajor and semiminor reference ellipsoid axes, γ_e and γ_p are derived theoretical gravity constants at the equator and poles, ϵ is eccentricity, and ϕ is geodetic latitude. The dependence of this formula on geodetic latitude will have consequences when closure errors arise in long level lines run mostly north to south compared to those that are run mostly east to west. For the International Gravity Formula 1930 (Int 30), Eq. 9.4 has the form (Blakely 1995, p. 135)

$$\gamma = 9.780\,46(1 + 0.005\,2884 \sin^2 \phi - 0.000\,0059 \sin^2 2\phi) \qquad (9.5)$$

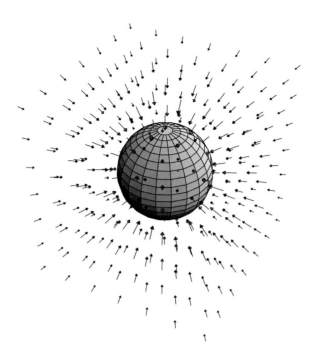

Figure 9.3: The gravitational force field of a spherical, homogeneous, nonrotating Earth. Note that the magnitude of the force decreases with separation from the Earth.

GRS 80 and WGS 84 are the most modern geodetic reference systems. Blakely (1995, p. 136), gives the closed-form formula for WGS 84 normal gravity as

$$g_0 = 9.780\,326\,7714 \frac{1 + 0.001\,931\,851\,386\,39 \sin^2 \phi}{\sqrt{1 - 0.006\,694\,379\,990\,13 \sin^2 \phi}} \tag{9.6}$$

Figure 9.4 shows a plot of the difference between Eq. 9.5 and Eq. 9.6. The older model has a larger value throughout and has, in the worst case, with the maximum difference being $0.000\,163\,229$ m/s^2 (i.e., about 16 mgal) at the equator. Gravity and gravitational acceleration are measured in gallileos (gal), which is 1 cm per second per second. The milligal (mgal) is 0.001 gal.

A normal potential field U gives rise to normal gravity. The Earth's actual potential field is denoted by W. Since the actual potential field is more irregular than the normal field, thinking of W as the "bumpier" of the two might help with remembering which is which.

9.2.3 Level surfaces

Equation 9.2 suggests that height might be inferred by measuring gravitational force because Eq. 9.2 can be solved for the magnitude of $\mathbf{r}$, which would be a height measured from the Earth's center of gravity. At first, this approach might seem to hold promise because the acceleration due to gravity can be detected with instruments that carefully measure the acceleration of a standard mass, either as a pendulum or free falling (Faller and Vitouchkine 2003). It seems such a strategy would deduce height in a way that stems from the physics that give rise to water's downhill motion and, therefore, would capture the underlying concept behind height very well.

9.2. PHYSICS

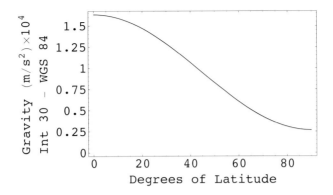

Figure 9.4: Difference in normal gravity between the International Gravity Formula 1930 (Eq. 9.5) and WGS 84. The values on the ordinate are given as 10 000 times the actual difference for clarity.

Regrettably, surfaces of constant gravity force (equiforce surfaces) are not necessarily level. Figure 9.5 shows the gravitational force field generated by two point-unit masses located at (0,1) and (0,-1). Note the lines of symmetry along the x- and y-axes. All forces for places on the x-axis are parallel to the axis and directed towards (0,0). Above or below the x-axis, all force lines ultimately lead to the mass also located on that side. Figure 9.6 shows a plot of the magnitudes of the vectors in Fig. 9.5. Note the local maxima around $x = \pm 1$ and the local minima at the origin. Figure 9.7 is a plot of the "northeast" corner of the force vectors superimposed on top of an isoforce plot of their magnitudes (the lines in Fig. 9.6 are places having the same magnitude of gravitational force). Note that the vectors are not perpendicular to the isolines. If one were to place a drop of water anywhere in the space represented by the figure, the water would follow the vectors to a peak, following and crossing isoforce lines along the way. This makes no sense if isoforce lines correspond to level surfaces. Level surfaces are defined with gravity potential energy.

9.2.4 Gravity potential energy

Work is what happens when a force is applied to an object causing it to move. Work plays a direct role in the definition of the geoid because an object's gravity potential energy changes when work causes the object to move with or against the force of gravity. This makes the concept of work fundamental to the definition of the geoid.

Work is a scalar quantity measured in joules (J), which are obtained by multiplying distance squared by mass per second squared. Only the force applied in the direction of motion contributes to the work done on the object.

Suppose an object moves in a straight line. With a constant force denoted by $\mathbf{F}$ and the displacement of the object denoted by a vector $\mathbf{s}$, the work done on the object is $w = \mathbf{F} \cdot \mathbf{s}$ (from Eq. A.1). This expression would be correct even if $\mathbf{F}$ was not directed exactly along the path of motion, because the inner product extracts from $\mathbf{F}$ only that portion that is directed parallel to $\mathbf{s}$. Of course, in general, force can vary with position, and the path of motion might not be a straight line. Let C denote a curve that has been parameterized by arc length s, meaning that $\mathbf{p} = C(s)$ is a point on C that is s units from C's starting point. Let $\hat{\mathbf{t}}(s)$ denote a unit vector tangent to C at s. Because we want to allow force to vary along C, we adopt a notation indicating that the force

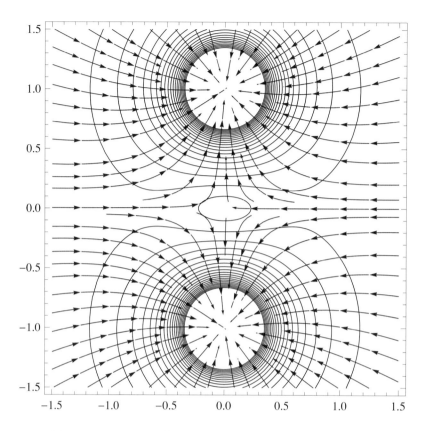

Figure 9.5: Force field created by two point masses.

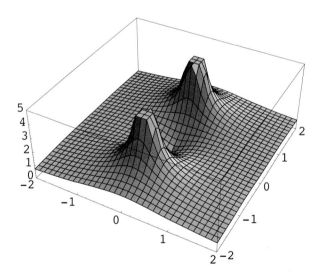

Figure 9.6: Magnitude of the force field created by two point masses.

9.2. PHYSICS

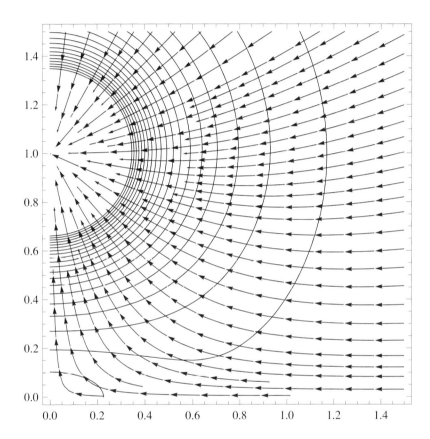

Figure 9.7: The force field vectors shown with the isoforce lines of the field.

is a function of position $\mathbf{F}(s)$. Then, the work expended by the application of a possibly varying force along a possibly curving path C from s_0 to s_1 is

$$w = \int_{s_0}^{s_1} \mathbf{F}(s) \cdot \hat{\mathbf{t}}(s) \mathrm{d}s \tag{9.7}$$

Equation 9.7 is general, and we can apply it as we turn our attention to motion within a gravity force field. Suppose we were to move some object in the presence of a gravity force field. What would be the effect? Let us first suppose that we move the object on a level surface, which implies that the direction of the gravity force vector is everywhere normal to that surface and, thus, perpendicular to $\hat{\mathbf{t}}(s)$ as well. By this assumption $\mathbf{F}_g$ is perpendicular to $\hat{\mathbf{t}}(s)$, so $\mathbf{F}_g$ plays no part in the work being done because $\mathbf{F}_g(s) \cdot \hat{\mathbf{t}}(s) = 0$. Therefore, moving an object over a level surface in a gravity field is identical to moving it in the absence of the field altogether, as far as the work done against gravity is concerned.

Now, suppose that gravity force is not everywhere normal to the direction of motion. Equation 9.7 implies that either more or less work will be needed due to the force of gravity, depending on whether the motion is against or with gravity, respectively. The gravity force will be accounted for by simply adding it to the force we apply; the distinction between them is not relevant to the object's motion. We can use superposition to separate the work done in the same direction as gravity from the work done to move laterally through the gravity field; they are orthogonal. Based on vector calculus, the work done by gravity on a moving body does not depend on the path of motion, apart from the starting and ending points. This is a consequence of gravity being a conservative field (Schey 1992; Blakely 1995). As a result, the work integral along the curve defining the path of motion can be simplified to consider work only in the direction of gravity. The path along which gravity acts is called a **plumb line** and, over short distances, can be considered to be a straight line, although plumb lines are not straight in general (see Fig. 9.7). Therefore, from Eq. 9.7, the work needed to move some object vertically through a gravity field is given by

$$w = \int_{h_0}^{h_1} \mathbf{F}_g(h) \cdot \hat{\mathbf{t}}(h) \mathrm{d}h \tag{9.8}$$

where h is height (distance along the plumb line), and $\hat{\mathbf{t}}(h)$ is the direction of gravity. However, $\mathbf{F}_g(h)$ is always parallel to $\hat{\mathbf{t}}(h)$, so $\mathbf{F}_g(h) \cdot \hat{\mathbf{t}}(h) = \pm F_g(h)$, depending on whether the motion is with or against gravity. If $\mathbf{F}_g(h)$ is constant, Eq. 9.8 can be simplified as

$$\begin{aligned} w &= \int_{h_0}^{h_1} \mathbf{F}_g(h) \cdot \hat{\mathbf{t}}(h) \mathrm{d}h \\ &= \int_{h_0}^{h_1} m\, \mathbf{E}_g(h) \cdot \hat{\mathbf{t}}(h) \mathrm{d}h && \text{(using Eq. 9.2)} \\ &= m\, \mathbf{E}_g(h) \int_{h_0}^{h_1} \mathrm{d}h && \text{(assuming } E_g \text{ is constant} = g\text{)} \\ &= m\, g\, \delta h \end{aligned} \tag{9.9}$$

where we denote the assumed constant magnitude of gravitational acceleration at the Earth's surface by g, as is customary. The quantity mgh is called **potential energy**. Equation 9.9 implies that releasing potential energy will do work if the object moves along gravity force lines. The linear dependence of Eq. 9.9 on height (h) is a key concept.

9.3 The Geoid

The Geodetic Glossary (NGS 2009) defines the **geoid** as "The equipotential surface of the Earth's gravity field which best fits, in a least squares sense, global mean sea level." The geoid is also called the "figure of the Earth." Shalowitz (1938, p. 10) writes, "The true figure of the Earth, as distinguished from its topographic surface, is taken to be that surface which is everywhere perpendicular to the direction of the force of gravity and which coincides with the mean surface of the oceans." The direction of gravity varies in a complicated way from place to place. Local vertical remains perpendicular to the undulating geoid, whereas local normal remains perpendicular to an ellipsoid reference surface. Recall that the angular difference of these two is the deflection of the vertical (see section 7.1.2).

Although Eq. 9.9 indicates a fundamental relationship between work and potential energy, we do not use this relationship directly because measuring work to find potential is not convenient. We rely on a direct relationship between the Earth's potential field and its gravity field:

$$\mathbf{E}_g = \nabla W \tag{9.10}$$

where W is the Earth's actual potential field, and ∇ is the gradient operator.[1] Written out in Cartesian coordinates, Eq. 9.10 becomes

$$\mathbf{E}_g = \frac{\partial W}{\partial X}\hat{\imath} + \frac{\partial W}{\partial Y}\hat{\jmath} + \frac{\partial W}{\partial Z}\hat{k}$$

where $\hat{\imath}, \hat{\jmath}$, and $\hat{k}$ are unit vectors in the X, Y, and Z directions, respectively. In spherical coordinates, Eq. 9.10 becomes

$$\mathbf{E}_g = \frac{\partial W}{\partial r}\hat{\mathbf{r}} \tag{9.11}$$

Equation 9.10 says that the gravity field is the gradient of the potential field (Blakely 1995; Heiskanen and Moritz 1967; Ramsey 1981; Vaníček and Krakiwsky 1996; Torge 1997; Hofmann-Wellenhof and Moritz 2005). Although Eq. 9.10 can be proven easily (Heiskanen and Moritz 1967, p. 2), the meaning of the equation may seem illusive.

Let us consider for a moment the question of why air bubbles rise towards the surface of the water? The usual answer is that air is lighter than water. This is surely true, but $\mathbf{F} = m\,\mathbf{a}$, so if bubbles are moving, then a force must be involved. Consider a bubble immersed in a water column (see Fig. 9.8). The pressure exerted by a column of water increases nearly linearly with depth (because water is nearly incompressible). The water exerts a force inwards on the bubble from all directions. If the forces were balanced, no motion would occur. It would be like a rope in a tug-of-war in which both teams are equally matched. Both teams are pulling the rope, but the rope is not moving, because equal and opposite forces cause no motion.

However, the bubble has finite height: the top of the bubble is at a shallower depth than the bottom of the bubble. Therefore, the pressure at the top of the bubble is less than the pressure at the bottom. This pressure gradient creates an excess of force from below that drives the bubble upwards. The difference in magnitude between any two lines of pressure is the gradient of the force

---

[1]Other authors write Eq. 9.10 as $\mathbf{E}_g = -\nabla W$, but the choice of the negative sign is essentially one of perspective. If the negative sign is included, the equation describes work done to overcome gravity. Without the negative sign, Eq. 9.10 follows directly from Eq. 9.2, in which the negative sign is necessary to capture the attraction of gravitational force.

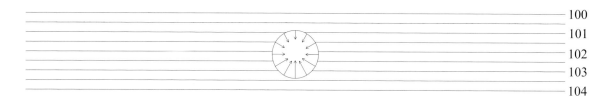

Figure 9.8: Force experienced by a bubble due to water pressure. Horizontal lines indicate surfaces of constant pressure, with sample values indicated on the side.

field; it is the potential energy of the force field. The situation with gravity is exactly analogous to the situation with water pressure. Any surface below the water at which the pressure is constant might be called an "equipressure" surface. Any surface in or around the Earth upon which the gravity potential is constant is called a **gravity equipotential surface**. A surface in a gravity field is created by the difference between the gravity potentials of two infinitely close gravity equipotential surfaces.

By assuming a spherical, homogeneous, nonrotating Earth, we can derive its potential field acting on a unit mass from Eq. 9.11. Denoting $|\mathbf{r}|$ by r

$$\frac{\partial W}{\partial r}\hat{\mathbf{r}} = \mathbf{E}_g$$
$$\mathrm{d}W = \frac{\mathbf{E}_g \mathrm{d}r}{\hat{\mathbf{r}}}$$
$$\int \mathrm{d}W = -\int \frac{GM}{r^2}\mathrm{d}r$$
$$W = \frac{GM}{r} + c \tag{9.12}$$

The constant of integration in Eq. 9.12 can be chosen so that zero potential resides either infinitely far away or at the center of M. If we choose the former convention, potential increases in the direction that gravity force vectors point, and the absolute potential of an object of mass m located a distance h from M is

$$\begin{aligned} W &= -\int_\infty^h \frac{GMm}{r^2}\mathrm{d}r \\ &= \frac{GMm}{r}\bigg|_\infty^h \\ &= \frac{GMm}{h} - \frac{GMm}{\infty} \\ &= \frac{GMm}{h} \end{aligned} \tag{9.13}$$

We now reconsider the definition of the geoid. According to Eq. 9.12, the geoid is associated with some particular value of W, and furthermore, if the Earth were spherical, homogeneous, and not rotating, the geoid would be located at some constant distance from the Earth's center of gravity. However, none of these assumptions are correct, so the geoid occurs at various distances from the Earth's center – it undulates.

9.3. THE GEOID

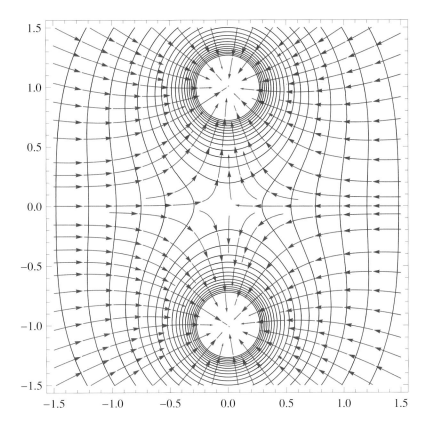

Figure 9.9: Gravitational force stream lines and the magnitude of the potential field created by two point masses.

One can prove mathematically that $\mathbf{E}_g$ is perpendicular to W, but this can also be illustrated visually. Figures 9.9 and 9.10 show the force vectors from Fig. 9.7 superimposed over the potential field computed using Eq. 9.12 instead of the magnitude of the force field. Notice that the vectors are perpendicular to the isopotential lines. Water would not flow along the isopotential lines; only across them. In three dimensions, the isopotential lines would be equipotential surfaces forming convex shells around the Earth. The geoid is one such shell, the shell that nominally coincides with mean sea level.

Equation 9.12 implies that the equipotential surfaces of a spherical, homogeneous, nonrotating mass would be concentric, spherical shells – much like layers of an onion. In a relatively small region near the surface of an Earth-sized sphere, the equipotential surfaces would almost be parallel planes. Figure 9.11 on the following page shows a cross section with gravitational force lines on the left and geopotential on the right. A very slight curvature is visible.

Now, suppose a point mass roughly equal to that of Mt. Everest positioned slightly below the surface is added to the sphere. The resulting gravity force field and isopotential lines are shown in Fig. 9.12. The deflection of the vertical is very apparent. Note how the shape of the isopotential lines run more or less horizontally, and how they bulge up over the mountain. This illustrates how equipotential surfaces roughly follow the topographic shape of the Earth in that they bow up over mountains and dip down into valleys. Also, any particular one of the geopotential lines

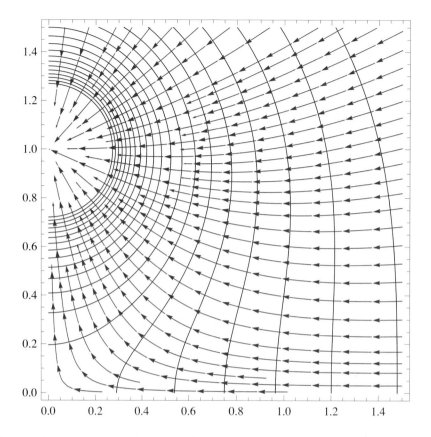

Figure 9.10: Upper right corner of Fig. 9.9. The stream lines are perpendicular to the isolines thus illustrating that equipotential surfaces are level.

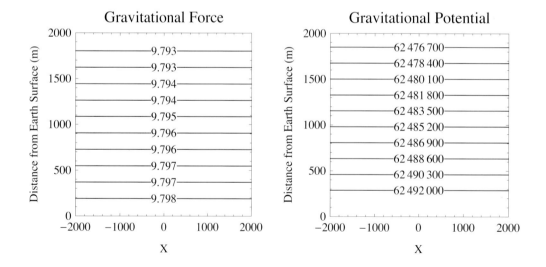

Figure 9.11: Gravitational force and geopotential isolines created at the surface of a uniform, nonrotating Earth.

9.4. GEOID HEIGHTS

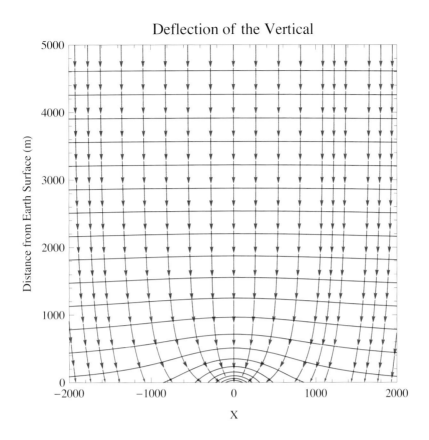

Figure 9.12: Effect of topography on the geoid: gravity force vectors and isopotential lines created at the Earth's surface by a point with a mass roughly equal to that of Mt. Everest. The gravity force vectors are plumb lines.

in Fig. 9.12 could be thought of as representing the surface of the ocean above an underwater seamount. Water piles up over the top of subsurface topography to exactly the degree that the weight of the additional water exactly balances the excess of gravitation caused by the seamount. Thus, one can indirectly observe seafloor topography by measuring the departure of the ocean's surface from nominal gravity (Hall 1992). The geoid, of course, surrounds the Earth, and Fig. 9.13 on page 187 shows the ellipsoid height of the geoid (geoid undulations) referred to the WGS 84 reference ellipsoid as placed by WGS 84(G1150), as computed by the Earth Gravity Model 2008 (EGM2008) on a $2.5' \times 2.5'$ global grid (Pavlis et al. 2008).

9.4 Geoid Heights

Reference ellipsoids that were custom fit to a particular region of the world are called **local reference ellipsoids**. Local ellipsoids do not provide a vertical datum in the ordinary sense, nor are they used as such. Local ellipsoids are essentially mathematical fictions that enable the conversion between geocentric, geodetic, and map projection coordinate systems in a rigorous way and, thus, provide part of the foundation of horizontal geodetic datums, but nothing more. According to

Fischer (2004), "O'Keefe [2] tried to explain to me that conventional geodesy used the ellipsoid only as a mathematical computation device, a set of tables to be consulted during processing, without the slightest thought of a third dimension."

In contrast to local ellipsoids that were the product of triangulation networks, globally applicable reference ellipsoids have been created using a variety of space geodesy techniques such as very long baseline interferometry (VLBI), satellite ranging, and Doppler observations of artificial satellites (Moritz 2000), along with various astronomical and gravitational measurements. Very long baseline interferometry and satellite observations permit high-accuracy baseline measurement between stations separated by oceans. Consequently, these ellipsoids model the Earth globally; they are not fitted to a particular local region.

WGS 84 and GRS 80 are not reference ellipsoids. They are geodetic reference systems, which provide the parameters necessary for constructing geodetic reference frames and datums (NIMA 1997). The WGS 84 and GRS 80 provide reference ellipsoids, which are best-fitting models of the geoid in a least-squares sense. According to Moritz (2000, p. 128),

> The Geodetic Reference System 1980 has been adopted at the XVII General Assembly of the IUGG in Canberra, December 1979, by means of the following: ... recognizing that the Geodetic Reference System 1967 ... no longer represents the size, shape, and gravity field of the Earth to an accuracy adequate for many geodetic, geophysical, astronomical and hydrographic applications and considering that more appropriate values are now available, [the International Association of Geodesy] recommends ... that the Geodetic Reference System 1967 be replaced by a new Geodetic Reference System 1980, also based on the theory of the geocentric equipotential ellipsoid, defined by the following constants:
>
> - Equatorial radius of the Earth: $a = 6\,378\,137$ m;
> - Geocentric gravitational constant of the Earth (including the atmosphere): $GM = 3\,986\,005 \times 10^8 \, \text{m}^3 \, \text{s}^{-2}$;
> - Dynamical form factor of the Earth, excluding the permanent tidal deformation: $J_2 = 108\,263 \times 10^{-8}$; and
> - Angular velocity of the Earth: $\omega = 7\,292\,115 \times 10^{-11} \, \text{rad s}^{-1}$.

Clearly, equipotential ellipsoid models of the Earth are a significant logical departure from local ellipsoids. Local ellipsoids are purely geometric, whereas equipotential ellipsoids also concern gravity. The GRS 80 reference ellipsoid is called an "equipotential ellipsoid" (Moritz 2000) and, using equipotential theory together with the defining constants listed above, one derives the ellipsoid's flattening rather than measuring it geometrically.

The geoid has a more complex surface than an equipotential ellipsoid, which can be completely described by just the four parameters listed above. The geoid's shape is strongly influenced by the topographic surface of the Earth. In Fig. 9.13, the geoid appears "bumpy," with mountains and valleys. Figure 9.13 shows the ellipsoid height of the geoid as estimated by EGM2008. Such heights (the ellipsoid height of a place on the geoid) are called **geoid heights**.

---

[2] John O'Keefe was the head of geodetic research at the Army Map Service.

9.4. GEOID HEIGHTS

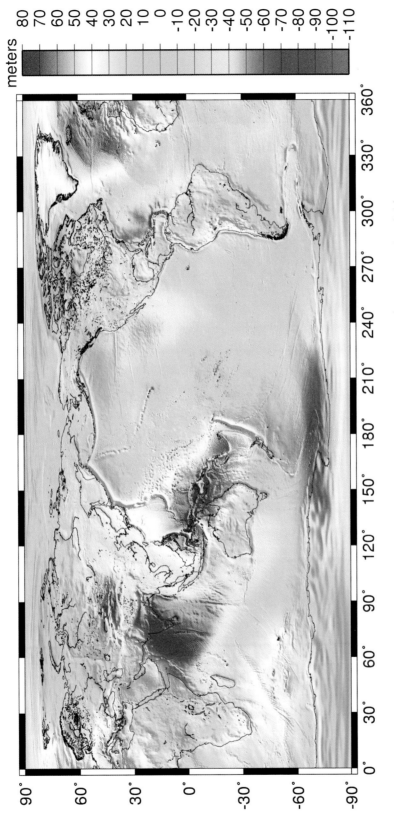

Figure 9.13: Geoid undulations from EGM2008 (courtesy of NGA).

9.5 Geopotential Numbers

The geoid is usually considered the proper surface from which to reckon geodetic heights because it honors the flow of water and nominally corresponds to mean sea level. Sea level itself does not exactly match the geoid, as explained above. Therefore, using the geoid as the level reference surface of a usable vertical datum is currently not possible using mean sea level measurements. Ideally, one would measure potential directly in some fashion analogous to measuring gravity acceleration directly, and the resulting number would be a **geopotential number**, which is the potential of the Earth's gravity field at any point in space. Using geopotential numbers as heights is appealing for several reasons:

- Geopotential defines hydraulic head. Therefore, if two points are at the same geopotential number, water will not flow between them due to gravity alone. Conversely, if two points are not at the same geopotential number, gravity will cause the water to flow between them if the waterway is unobstructed (ignoring friction).

- Geopotential decreases linearly with distance from the center of the Earth (Eq. 9.12). This makes it a natural measure of distance.

- Geopotential does not depend on the path taken from the Earth's center to the point of interest. This makes a geopotential number stable.

- Since the magnitude of a geopotential number is less important than the relative values between two places, geopotential numbers can be scaled to any desirable values, and the geoid can be defined to have a geopotential number of zero.

Equation 9.13 might seem to imply that height can be determined by measuring absolute potential. Regrettably, potential cannot be measured directly because the manifestation of potential (gravitational force) is created by potential differences (the gradient of potential, to be precise) and not the potential itself. Any two places separated by the same change in geopotential have gravitational forces of the same magnitude. In light of this, one might ask how images of the geoid are created. Figure 9.13 is the result of a sophisticated mathematical model with theory based on Stokes' formula (Heiskanen and Moritz 1967, p. 94, Eq. 2-163b):

$$N = \frac{R}{4\pi G} \int_\sigma \Delta g \, S(\psi) \mathrm{d}\sigma \qquad (9.14)$$

where N is geoid height at a point of interest, R is mean radius of the Earth, G is the universal gravitational constant, Δg is the reduced, observed gravity measurements around the Earth, ψ is spherical distance from each surface element $\mathrm{d}\sigma$ to the point of interest, and $S(\psi)$ is Stokes' function (Heiskanen and Moritz 1967, p. 94, Eq. 2-164):

$$S(\psi) = \frac{1}{\sin(\psi/2)} - 6\sin(\psi/2) + 1 - 5\cos\psi - 3\cos\psi \ln(\sin(\psi/2) + \sin^2(\psi/2))$$

The model is calibrated with, and has boundary conditions provided by, reduced gravity measurements taken in the field – the Δg's in Eq. 9.14. These measurements together with Stokes' formula permit the deduction of the potential field that must have given rise to the observed gravity measurements.

In spite of their natural suitability, geopotential numbers are not practical to use as heights because surveyors cannot easily measure them in the field. Also, geopotential numbers have units of energy per unit mass, not length. Most surveyors would probably object to using heights that don't have length units, as well. Geopotential numbers are, however, the essence of what the word *height* really means.

9.6 Your Turn

Problem 9.1. Contour maps show lines of constant orthometric height. Draw a contour map of the geoid.

Problem 9.2. Why is the force field shown in Fig. 9.5 on page 178 zero at the origin?

Problem 9.3. Draw a graph showing the gravitational force field's magnitude created by an Earth-sized, homogeneous, nonrotating sphere from the sphere's center to its surface. (Hint: use arguments like the ones we used to explain the motion of bubbles in a water column.)

Chapter 10

Height Systems

The word *height* is based on two general visions: a geometric separation and hydraulic head. These visions have led to many formulations of different types of heights: orthometric heights, purely geometric heights, and heights that are neither. They all have strengths and weaknesses, and this has given rise to different height systems.

10.1 Types of Heights

10.1.1 Orthometric heights

Heiskanen and Moritz (1967, p. 172) write, "Orthometric heights are the natural 'heights above sea level,' that is, heights above the geoid. They thus have an unequalled geometrical and physical significance." The Geodetic Glossary (NGS 2009) defines **orthometric height** as "The distance between the geoid and a point measured along the plumb line and taken positive upward from the geoid," with **plumb line** defined as "A line perpendicular to all equipotential surfaces of the Earth's gravity field that intersect with it." Orthometric height can be defined more formally as the length of the integral curve of the vector field representing the acceleration due to gravity at every point measured from a point of interest to the geoid (Craig Rollins, personal communication). In one sense, orthometric height is geometric: it is the length of a segment of a particular curve (a plumb line). However, that curve depends on gravity in two ways. First, the curve begins at the geoid. Second, plumb lines remain everywhere perpendicular to equipotential surfaces through which they pass, so the shape of the curve is determined by the orientation of the equipotential surfaces. Therefore, orthometric heights are closely related to gravity in addition to having a geometric quality.

How are orthometric heights related to geopotential? From Eq. 10.5, $g = -\delta W/\delta H$. Taking differentials instead of finite differences and rearranging, we get $dW = -g\, dH$. Recall that a geopotential number is the difference in potential between the geoid W_0 and a point of interest **A**:

$C_\mathbf{A} = W_0 - W_\mathbf{A}$, so

$$\int_{W_0}^{W_\mathbf{A}} \mathrm{d}W = -\int_0^{H_\mathbf{A}} g\,\mathrm{d}H$$

$$W_\mathbf{A} - W_0 = -\int_0^{H_\mathbf{A}} g\,\mathrm{d}H$$

$$W_0 - W_\mathbf{A} = \int_0^{H_\mathbf{A}} g\,\mathrm{d}H$$

$$C_\mathbf{A} = \int_0^{H_\mathbf{A}} g\,\mathrm{d}H \tag{10.1}$$

with the understanding that gravity acceleration g is not a constant. Equation 10.1 can be used to derive the desired relationship:

$$C_\mathbf{A} = \bar{g}\,H_\mathbf{A} \tag{10.2}$$

meaning that a geopotential number is equal to an orthometric height multiplied by the average acceleration of gravity along the plumb line. As was argued in chapter 9, geopotential is single-valued, meaning the potential of any particular place is independent of the path taken to arrive there. Consequently, orthometric height is likewise single-valued, being a scaled value of a geopotential number.

If orthometric heights are single-valued, it is logical to inquire whether surfaces of constant orthometric height form equipotential surfaces. The answer to this is, unfortunately, no. If orthometric heights formed equipotential surfaces, two different places at the same orthometric height would be at the same potential. With this hypothesis, Eq. 10.2 requires that the average gravity along the plumb lines of these different places necessarily be equal. However, the acceleration of gravity depends on height, latitude, and the distribution of nearby masses; neither its magnitude nor its direction is constant. There is no reason that the average gravities would be equal, and in fact, they typically are not. Two points at the same orthometric height can have different gravity potential energies, so they would be on different equipotential surfaces and would be at different heights from the perspective of geopotential numbers.

Figure 10.1, which is essentially a three–dimensional rendering of Fig. 9.12 on page 185, shows an imaginary mountain and various equipotential surfaces. Figure 10.1.b shows the mountain with just one gravity equipotential surface. All points on a gravity equipotential surface are at the same gravity potential, so water would not flow along the intersection of the equipotential surface with the topography without external impetus. Nevertheless, the curve defined by the intersection of the gravity equipotential surface with the topography would not be drawn as a contour line on a topographic map because a **contour line** is "An imaginary line on the ground, all points of which are at the same *elevation* above or below a specified reference surface" (NGS 2009). This runs contrary to conventional wisdom that would define a contour line as the intersection of a horizontal plane with the topography. In Fig. 10.1.c and d, the equipotential surfaces undulate; they do not remain everywhere the same distance apart from each other, and they "pull up" through the mountains. Figure 10.1.d shows multiple surfaces, each having less curvature than the one below it as a consequence of increasing distance from the Earth.

Figure 10.2 is an enlargement of the foothill appearing on the right in Fig. 10.1.c. Suppose that the equipotential surface containing **A** and **D** is the geoid. Then the orthometric height of station **B** is the distance along its plumb line to the surface containing **A** and **D**; the same is true

10.1. TYPES OF HEIGHTS

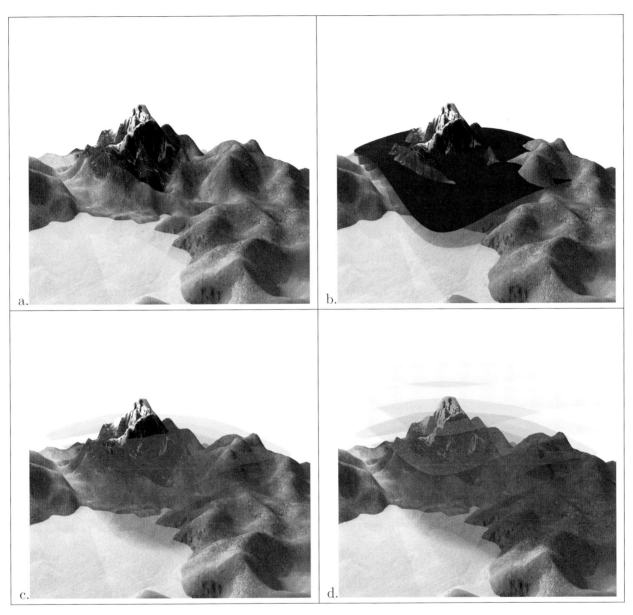

Figure 10.1: Four views of several geopotential surfaces around and through an imaginary mountain. (a) The mountain without any equipotential surfaces. (b) One equipotential surface is shown. The intersection of the surface and the ground is a line of constant gravity potential but not a contour line. (c) Two equipotential surfaces. The surfaces are not parallel, and they undulate through the terrain. (d) Many equipotential surfaces. The further a surface is from the Earth, the less curvature it has.

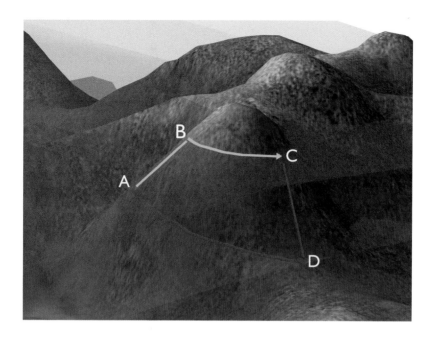

Figure 10.2: Two lines of levels illustrating why orthometric heights do not, in principle, produce closed leveling circuits.

for station **C**. Although neither **B**'s nor **C**'s plumb line is shown (both plumb lines are inside the mountain), one can see that the separation from **B** to the geoid is different than the separation from **C** to the geoid, even though **B** and **C** are on the same equipotential surface. Therefore, they have the same geopotential number but have different orthometric heights. This illustrates why orthometric heights are single-valued but do not create equipotential surfaces.

How are orthometric heights measured? Suppose an observed sequence of geometric height differences δv_i have been summed together for the total change in geometric height along a leveling section from station **A** to **B**: $\delta v_{\mathbf{AB}} = \sum_i \delta v_i$. We can denote the change in orthometric height from **A** to **B** as $\delta H_{\mathbf{AB}}$. Equation 10.2 requires knowing a geopotential number and the average acceleration of gravity along the plumb line, but neither of these is measurable. Fortunately, there is a relationship between a leveling difference and an orthometric height difference. A change in orthometric height equals a change in geometric height plus a correction factor known as the **orthometric correction** (Heiskanen and Moritz 1967, p. 167-168, Eq. 4-31 and 4-33):

$$\delta H_{\mathbf{AB}} = \delta v_{\mathbf{AB}} + OC_{\mathbf{AB}} \qquad (10.3)$$

where $OC_{\mathbf{AB}}$ is the orthometric correction. It is

$$OC_{\mathbf{AB}} = \sum_{\mathbf{A}}^{\mathbf{B}} \frac{g_i - \gamma_0}{\gamma_0} \delta v_i + \frac{\bar{g}_{\mathbf{A}} - \gamma_0}{\gamma_0} H_{\mathbf{A}} - \frac{\bar{g}_{\mathbf{B}} - \gamma_0}{\gamma_0} H_{\mathbf{B}} \qquad (10.4)$$

where g_i is the observed force of gravity at the observation stations; $\bar{g}_{\mathbf{A}}$ and $\bar{g}_{\mathbf{B}}$ are the average values of gravity along the plumb lines at **A** and **B**, respectively; and γ_0 is an arbitrary constant, which is often taken to be the value of normal gravity at a latitude of 45° (see section 9.2.2). (See

10.1. TYPES OF HEIGHTS

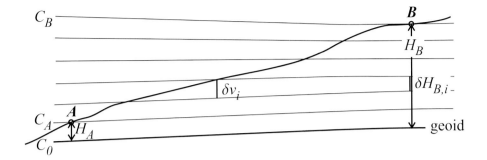

Figure 10.3: Comparison of differential-leveling height differences δv_i with orthometric height differences $\delta H_{B,i}$.

Hwang and Hsiao (2003) for an orthometric correction formula that does not depend on normal gravity.) Although Eq. 10.4 stipulates that gravity needs to be observed at every measuring station, Bomford (1962, p. 206) suggests that the observation stations need to be no closer than 2-3 km in level country but should be as close as 0.3 km in mountainous country. Others recommend observation station separations of 15 to 25 km in level country and 5 km in mountainous country (Strang van Hees 1992; Kao et al. 2000; Hwang and Hsiao 2003).

Practical applications of orthometric corrections have been described by Forsberg (1984), Strang van Hees (1992), Kao et al. (2000), Allister and Featherstone (2001), Hwang (2002), Brunner (2002), Hwang and Hsiao (2003), and Tenzer et al. (2005). The work described in these reports was undertaken by institutions with the resources to field gravimeters with accompanying surveying crews. Although progress has been made in developing portable gravimeters (Faller and Vitouchkine 2003), gravity measurements called for by Eq. 10.4 remain impractical for most surveyors. For First-Order leveling, the National Geodetic Survey has used corrections that depend solely on the geodetic latitude and normal gravity at the observation stations, thus avoiding the need to measure gravity (NGS 1981, p. 5–26), although if leveling is used to determine geopotential numbers (as in the NAVD 88 adjustment), orthometric corrections aren't used. In the United States, surveyors can use modeled gravity (GRAV) values found on NGS data sheets for the values of g_i. Here is an excerpt from the NGS data sheet for PID LX3030:

```
LX3030   LAPLACE CORR-          -2.38  (seconds)                   DEFLEC99
LX3030   ELLIP HEIGHT-         157.28  (meters)     (06/22/01)     GPS OBS
LX3030   GEOID HEIGHT-         -29.58  (meters)                    GEOID03
LX3030   DYNAMIC HT  -         186.779 (meters)      612.79 (feet) COMP
LX3030   MODELED GRAV-     980,277.0   (mgal)                      NAVD 88
```

Modeled gravity values can be obtained at any U.S. location, not just at a survey marker, using the NAVDGRAD program (NGS 2008). Modeled gravity values for any location worldwide can be obtained using EGM2008 (Pavlis et al. 2008).

Although $\bar{g}$ cannot be determined exactly, it can be estimated with either a free-air correction (Heiskanen and Moritz 1967, p. 163–164) or the Poincaré–Prey reduction (ibid., p. 165). The former depends on knowledge of normal gravity only by making assumptions regarding the mean curvature of the potential field outside of the Earth. Orthometric heights that depend upon this strategy are called **Helmert orthometric heights**. NGS publishes NAVD 88 Helmert orthometric heights. The Poincaré–Prey reduction, which requires a remove-reduce-restore operation, is more complicated and only improves the estimate slightly (ibid., p. 163–165).

In summary, orthometric heights

- are the embodiment of the concept of height above sea level;

- are single-valued by virtue of their relationship with geopotential numbers and, consequently, produce closed leveling circuits, in theory;

- do not define equipotential surfaces because the force of gravity is variable. This implies that water could, in principle, flow uphill, if uphill means going from a place with a lower orthometric height to a place with higher orthometric height. Although possible, this situation would require a steep gravity gradient in a location with relatively little topographic relief. This can occur in places where subterranean features substantially affect the local gravity field but have no expression on the Earth's surface; and

- are not directly measurable from their definition. Orthometric heights can be determined by observing geometric height differences derived from differential leveling and applying the orthometric correction, which requires surface gravity observations and an approximation of the average acceleration of gravity along the plumb line.

10.1.2 Uncorrected differential leveling

Uncorrected differential leveling is a process by which the geometric height difference along the vertical is transferred from a reference station to a forward station. Differential leveling does not, in general, produce orthometric heights. Suppose a leveling line connects two stations $\mathbf{A}$ and $\mathbf{B}$ defining a section as depicted in Fig. 10.3 on the preceding page (cf. Heiskanen and Moritz 1967, p. 161). If the two stations are far enough apart, the section will contain several leveling turning points, with δv_i being the vertical geometric separation between such points. Any two turning points are at two particular geopotential numbers, the difference between which is the potential gravity energy available to move water between them. This difference is hydraulic head. The vertical geometric separation of two equipotential surfaces along the plumb line for $\mathbf{B}$ is $\delta H_{\mathbf{B},i}$.

The two stations $\mathbf{A}$ and $\mathbf{B}$ have geopotential numbers $C_{\mathbf{A}}$ and $C_{\mathbf{B}}$ and orthometric heights $H_{\mathbf{A}}$ and $H_{\mathbf{B}}$, respectively. The geopotential surfaces are not parallel; they converge toward the right. Therefore, $\delta v_i \neq \delta H_{\mathbf{B},i}$. The height difference from $\mathbf{A}$ to $\mathbf{B}$ as determined by uncorrected differential leveling is the sum of the δv_i. Therefore, because $\delta v_i \neq \delta H_{\mathbf{B},i}$ and the orthometric height at $\mathbf{B}$ can be written as $H_{\mathbf{B}} = \sum_i \delta H_{\mathbf{B},i}$, it follows that $\sum_i \delta v_i \neq H_{\mathbf{B}}$.

In the bubble gedankenexperiment in section 9.3, the force moving the bubble was the result of a change in water pressure over a finite change in depth. By analogy, gravity force is the result of a change in gravity potential over a finite separation:

$$g = -\delta W/\delta H \qquad (10.5)$$

where g is gravity force, W is geopotential, and H is orthometric height. Equation 10.5 can be rearranged as $-\delta W = g\,\delta H$. Since δv_i and $\delta H_{\mathbf{B},i}$ are, by construction, across the same potential difference, $-\delta W = g\,\delta v_i = g'\,\delta H_{\mathbf{B},i}$, where g' is gravity force at the plumb line. Now, $\delta v_i \neq \delta H_{\mathbf{B},i}$ due to the nonparallelism of the equipotential surfaces, but δW is the same for both, so gravity

10.1. TYPES OF HEIGHTS

on the surface where the leveling took place must be different at the plumb line. This leads us to (Heiskanen and Moritz 1967, p. 161, Eq. 4-2)

$$\delta H_{\mathbf{B},i} = \frac{g}{g'}\delta v_i \neq \delta v_i \tag{10.6}$$

which indicates that differential-leveling height differences differ from orthometric height differences by the amount that surface gravity differs from gravity along the plumb line at that geopotential. An immediate consequence of this is that two different lines of levels starting and ending at the same station will, in general, provide different values for the height of final station. This is because the two lines will run through different topography and, consequently, geopotential surfaces with disparate separations. Therefore, in principle, uncorrected differential leveling heights are not single-valued, meaning the result you get depends on the route you took to get there.

In summary, heights derived from uncorrected differential leveling

- are readily observed by differential leveling;
- are not single-valued by failing to account for the variability in gravity;
- will not, in theory, produce closed leveling circuits; and
- do not define equipotential surfaces. Indeed, they do not define surfaces in the mathematical sense at all.

10.1.3 Ellipsoid heights and geoid heights

Ellipsoid heights are the straight-line distances normal to a reference ellipsoid produced away from (positive) (or into [negative]) the ellipsoid to the point of interest. Before GNSS positioning, it was practically impossible for anyone outside the geodesy community to determine an ellipsoid height. Now, GNSS receivers produce three-dimensional baselines (Meyer 2002) resulting in determinations of geodetic latitude, longitude, and ellipsoid height. Today, ellipsoid heights are commonplace.

Ellipsoid heights are seldom suitable surrogates for orthometric heights over large areas because ellipsoid heights do not undulate with the geoid (see section 9.4 on page 185 and Kumar 2005). Furthermore, nowhere in the conterminous United States is the geoid closer to a GRS 80-shaped ellipsoid centered at the NAD 83(CORS96) origin than about 2 m. Confusing an ellipsoid height with an orthometric height would result in a blunder of at least 2 m and could potentially lead to disastrous consequences. For example, using ellipsoid heights instead of orthometric heights, the height of an obstruction in the approach to an airport runway in New York City would be reported about 30 m lower than its orthometric value, which would cause a pilot to mistakenly believe the aircraft had 30 m more clearance than it does.

Deriving orthometric heights from ellipsoid heights is mathematically very simple. If plumb lines were straight and if they were normal to the reference ellipsoid, then

$$H = h - N \tag{10.7}$$

where H is orthometric height, N is geoid height, and h is ellipsoid height. However, plumb lines are curved and they are not normal to reference ellipsoids, in general. Therefore, we cannot use an equality relationship and must instead write

$$H \approx h - N \tag{10.8}$$

Although Eq. 10.8 is not exact, it is close enough for most practical purposes (Hein 1985; Zilkoski and Hothem 1989; Zilkoski 1990; Henning et al. 1998; Vaníček et al. 1999). For example, an extreme case of a 2-arc minute deflection of the vertical would introduce less than 2 mm of error in the orthometric height (Tenzer et al. 2005, p. 89).

In summary, ellipsoid heights

- are single-valued,
- do not use the geoid or any other physical gravity equipotential surface as their datum,
- do not define equipotential surfaces, and
- are readily determined using GNSS positioning.

10.1.4 Geopotential numbers and dynamic heights

Geopotential numbers C are defined on the basis of Eq. 9.8, (cf. Heiskanen and Moritz 1967, p. 162, Eq. 4-8), which gives the change in gravity potential per unit mass between a point on the geoid and another point of interest. The geopotential number for any place is the potential of the geoid W_0 minus the potential of that place W (recall that the potential decreases with distance away from the Earth, so this difference is a positive number). Geopotential numbers are given in geopotential units (g.p.u.), where 1 g.p.u. = 1 kgal meter = 1000 gal meter (Heiskanen and Moritz 1967, p. 162, Eq. 4-8). If gravity is assumed to be a constant 0.98 kgal, a geopotential number is approximately equal to $0.98\,H$, so geopotential numbers in g.p.u. are nearly equal to orthometric heights in meters. However, geopotential numbers have units of energy per unit of mass, not length, and are therefore an "unnatural" measure of height.

Geopotential numbers can be scaled by dividing by a gravity value, which will change their units from kgal meters to meters. Doing so results in a **dynamic height**:

$$H^{dyn} = C/\gamma_0 \qquad (10.9)$$

One reasonable choice for γ_0 is the value of normal gravity (Eq. 9.6) at some latitude, often taken to be 45°. Obviously, scaling geopotential numbers by a constant does not change their fundamental properties, so dynamic heights, like geopotential numbers, are single-valued, produce equipotential surfaces, and form closed leveling circuits. They are not, however, geometric like orthometric heights: two different places on the same equipotential surface have the same dynamic height but generally do not have the same orthometric height. Thus, dynamics heights are not "distances from the geoid."

Dynamic heights can be determined in a manner similar to that for orthometric heights: geometric height differences observed by differential leveling are added to a correction term that accounts for gravity:

$$\delta H^{dyn}_{\mathbf{AB}} = \delta v_{\mathbf{AB}} + DC_{\mathbf{AB}} \qquad (10.10)$$

where $\delta v_{\mathbf{AB}}$ is the total measured geometric height difference derived by differential leveling and $DC_{\mathbf{AB}}$ is the dynamic correction. The **dynamic correction** from station **A** to station **B** is given by Heiskanen and Moritz (1967, p. 163, Eq. 4-11) as

$$DC_{\mathbf{AB}} = \sum_{A}^{B} \frac{g_i - \gamma_0}{\gamma_0} \delta v_i \qquad (10.11)$$

10.1. TYPES OF HEIGHTS

where g_i is the (variable) force of gravity at each leveling observation station, γ_0 is normal gravity at 45°, and δv_i is the observed change in geometric height along each section of the leveling line.

$DC_{\mathbf{AB}}$ typically is a large value for inland leveling conducted far from the defining latitude. For example, suppose a surveyor in Albuquerque, New Mexico (at a latitude of around 35° N), begins a level line at the Route 66 bridge over the downtown railroad tracks at an elevation of 1510 m, and runs levels to the Four Hills subdivision at an elevation of 1720 m, a change in elevation of 210 m. From Eq. 10.11, $DC_{\mathbf{AB}} = \delta v(g - \gamma_0)/\gamma_0$. Taking $\gamma_0 = \gamma_{45°} = 980.62$ gal and $\gamma_{35°} = 979.734$ gal, we get $DC_{\mathbf{AB}} = 210$ m $(979.734\,\text{gal} - 980.62\,\text{gal})/980.62\,\text{gal} = -0.190$ m, a correction of roughly two parts in one thousand. This is a huge correction compared to any other correction applied in First-Order leveling with no obvious physical interpretation such as the refraction caused by the atmosphere. Surveyors would not likely embrace a height system that imposed such large corrections, as the corrections would often affect even lower-accuracy work. Nonetheless, dynamic heights are of practical use wherever water levels are needed, such as the Great Lakes and also along ocean shores, even if they are used far from the latitude of the normal gravity constant. The geoid is thought to be not more than a couple meters from the ocean surface, and therefore, shores have geopotential near to that of the geoid. Consequently, shores have dynamic heights near zero regardless of their distance from the defining latitude. Dynamic heights in the International Great Lakes Datum of 1985 were established by the Vertical Control-Water Levels Subcommittee under the Coordinating Committee on Great Lakes Basic Hydraulics and Hydrology Data (CCGLBHHD). Even so, for inland surveying, $DC_{\mathbf{AB}}$ can have a large value, on the order of several meters at the equator.

In summary, dynamic heights

- are geopotential numbers scaled by a constant to endow them with units of length;

- are not geometric distances;

- are single-valued by virtue of their relationship with geopotential numbers, and consequently, produce closed circuits, in theory;

- define equipotential surfaces; and

- are not measurable directly from their definition. Dynamic heights can be determined by observing differential-leveling-derived geometric height differences to which the dynamic correction is applied. The dynamic correction requires surface gravity observations and can be on the order of meters in places far inland and far from the latitude at which γ_0 was defined.

10.1.5 Normal heights

Of heights defined by geopotential (orthometric and dynamic), Heiskanen and Moritz (1967, p. 287) write,

> The advantage of this approach is that the geoid is a level surface, capable of simple definition in terms of the physically meaningful and geodetically important potential W. The geoid represents the most obvious mathematical formulation of a horizontal surface at mean sea level. This is why the use of the geoid simplifies geodetic problems and makes them accessible to geometrical intuition.

The disadvantage is that the potential W inside the Earth, and hence the geoid W = const., depends on [a detailed knowledge of the density of the Earth]... Therefore, in order to determine or to use the geoid, the density of the masses at every point between the geoid and the ground must be known, at least theoretically. This is clearly impossible, and therefore some assumptions concerning the density must be made, which is unsatisfactory theoretically, even though the practical influence of these assumptions is usually very small.

These issues led Molodensky in 1945 to formulate a new type of height, normal height, which is based on the assumption that the Earth's gravity field is normal, meaning the actual gravity potential equals normal gravity potential (Molodensky 1945). With this postulate the "physical surface of the earth can be determined from geodetic measurements alone, without using the density of the earth's crust" (Heiskanen and Moritz 1967, p. 288). This conceptualization allowed determining heights by a fully rigorous method, a method without assumptions. The price, however, was that "This requires that the concept of the geoid be abandoned. The mathematical formulation becomes more abstract and more difficult" (ibid.). Normal heights are defined by

$$C = \int_0^{H^*} \gamma \, dH^* \quad (10.12)$$

and

$$C = \bar{\gamma} \, H^* \quad (10.13)$$

where H^* is normal height and γ is normal gravity. These formulæ have identical forms to those for orthometric height (Eq. 10.1 and 10.2) but their meaning is completely different. First, the zero used as the lower integral bound is not the geoid; it is a reference ellipsoid. Consequently, normal heights depend upon the choice of reference ellipsoid and frame. Second, normal gravity is an analytical function, so its average may be computed in closed form; no gravity observations are required. Third, by definition, a normal height H^* is that ellipsoid height where the normal gravity potential equals the actual geopotential of the point of interest. As Heiskanen and Moritz (1967, p. 170) put it, "...but since the potential of the earth is evidently not normal, what does all this mean?"

Like orthometric and dynamic heights, normal heights can be determined by applying a correction to geometrical height differences observed by differential leveling. The **normal correction** has the same structure as that for orthometric correction:

$$NC_{\mathbf{AB}} = \sum_{\mathbf{A}}^{\mathbf{B}} \frac{g_i - \gamma_0}{\gamma_0} \delta v_i + \frac{\bar{\gamma}_{\mathbf{A}} - \gamma_0}{\gamma_0} H_{\mathbf{A}}^* - \frac{\bar{\gamma}_{\mathbf{B}} - \gamma_0}{\gamma_0} H_{\mathbf{B}}^* \quad (10.14)$$

with $\bar{\gamma}_A$ and $\bar{\gamma}_B$ being the average normal gravity at **A** and **B**, respectively, and other terms defined as for Eq. 10.4. Normal corrections also depend upon gravity observations g_i but do not require assumptions regarding average gravity within the Earth. Therefore, they are rigorous; all the necessary quantities can be calculated or directly observed. Like orthometric heights, they do not form equipotential surfaces (because of normal gravity's dependence on latitude; recall that dynamic heights scale geopotential simply by a constant whereas orthometric and normal heights' scale factors vary with location). Like orthometric heights, normal heights are single valued and give rise to closed leveling circuits. Geometrically, they represent the distance from the ellipsoid up to a surface known as the telluroid (Heiskanen and Moritz 1967).

In summary, normal heights

Table 10.1: Comparison of height systems.

Height System	Single-valued	Defines level surfaces	No misclosure	Small correction	Physically meaningful	Rigorous implementation
Uncorrected differential leveling	No	No	No	NA[a]	Yes	Yes
Helmert orthometric	Yes	No	Yes	Yes	Yes	No
Ellipsoidal	Yes	No	Yes	NA	Yes	Yes
Dynamic	Yes	Yes	Yes	No	Yes	Yes
Normal	Yes	No	Yes	Yes	No	Yes

[a] NA, not applicable.

- are geometric distances, being ellipsoid heights, but not to the point of interest;

- are single-valued and, consequently, produce closed-circuits, in theory;

- do not define equipotential surfaces; and

- are not measurable directly from their definition. Normal heights can be determined by observing differential-leveling-derived geometric height differences to which the normal correction is applied. The normal correction requires surface gravity observations only and, therefore, can be determined without approximations.

10.2 Height Systems

A height system is a mechanism by which height values can be assigned to places of interest. Hipkin (2002) suggests the following:
(i) Height must be single-valued.
(ii) A surface of constant height must also be a level (equipotential) surface.
Heiskanen and Moritz (1967, p. 173) suggest the following:
(i) Misclosures must be eliminated.
(ii) Corrections to the measured heights must be as small as possible.
Hipkin's first criterion is equivalent to Heiskanen and Moritz's first criterion: if heights are single-valued, then leveling circuits will be closed and vice versa. The second criteria form the basis of two different philosophies about what is considered important for heights. Requiring that a surface of constant height be equipotential requires that the heights be scaled geopotential numbers and excludes orthometric and normal heights. Conversely, requiring the measurement corrections to be as small as possible precludes the former, at least from a global point of view, because dynamic height scale factors can be large. No height meets all these criteria. This has given rise to the use of (Helmert) orthometric heights in the United States, dynamic heights in Canada, and normal heights in Europe (Ihde and Augath 2000). Table 10.1 provides a comparison of these height systems (compare with Table 1. in Dennis and Featherstone (2002)).

The geoid is widely accepted as the proper datum for a height system, although this perspective has challengers (Hipkin 2002). Conceptually, the geoid is the natural choice for a vertical reference system, and until recently, its surrogate, mean sea level, was the object from which the geoid was realized. However, no modern vertical reference system, in fact, uses the geoid as its datum primarily because the geoid is difficult to realize.

The geoid is not directly realizable from mean sea level because the mean sea surface is not a level surface and dynamic topography is not directly observable (see section 9.1 on page 171). Hipkin (2002, p. 376) states that the "...nineteenth century approach to establishing a global vertical datum supposed that mean sea level could bridge regions not connectible by leveling. The geoid was formalized into the equipotential [surface] best fitting mean sea level and, for more than a century, the concepts of mean sea level, the geoid, and the leveling datum were used synonymously." We now know this use of "geoid" for "mean sea level" and vice versa to be incorrect because the mean sea surface is not an equipotential surface. Therefore, the mean sea surface is questionable as a vertical reference surface. Furthermore, Hipkin argues that measuring changing sea levels is one of the most important contributions that geodesy is making today. For this particular application, it does not make sense to continually adjust the vertical datum to stay at mean sea level and, thus, eliminate the phenomena trying to be observed. Conversely, chart makers, surveyors, and cartographers who define floodplains and subsidence zones usually require that the vertical datum reflect changes in sea level to ensure their products are up-to-date and not misleading.

It is not straightforward to produce a globally acceptable definition of the geoid. Smith (1998, p. 17) writes, "The Earth's gravity potential field contains infinitely many level surfaces...The geoid is one such surface with a particular potential value, W_0." W_0 is a fundamental geodetic parameter (Burša 1995; Groten 2004), and its value has been estimated by using sea surface topography models (also called dynamic ocean topography models) and spherical harmonic expansions of satellite altimetry data (Burša 1969; Burša 1994; Nesvorny and Sima 1994; Burša et al. 1997; Burša et al. 1999), as well as GNSS observations on orthometric bench marks (Grafarend and Ardalan 1997). Research conducted in a joint effort between NGS, the National Aeronautics and Space Administration Goddard Flight Center, and the Naval Research Laboratory is attempting to model the geoid by coupling sea surface topography model results with airborne gravimetry and Light Detection And Ranging (LIDAR) measurements in a manner similar to the space-based altimetry efforts. If successful, this work will result in another solution to the problem of determining W_0 with particular focus on the coastal regions of the United States (Smith and Roman 2001, p. 472). NGS and NGA are also examining Earth gravity models (EGMs) derived from the satellite-based Gravity Recovery and Climate Experiment (GRACE) (Tapley et al. 2004) and Gravity Field and Steady-State Ocean Circulation Explorer (GOCE) data (Rebhan et al. 2000) to establish higher confidence in the long wavelengths in EGMs (i.e., macroscopic scale features in the geoid model). Aerogravity data are being collected to bridge the gaps at the shorelines between terrestrial data and the deep ocean and altimeter-implied gravity anomalies. EGMs and aerogravity data are being used to cross-check each other and the existing terrestrial data. The most current Earth gravity model is EGM2008.

Experts do not agree on how W_0 should be chosen. Smith (1998) suggested W_0 could be chosen at least two ways: pick a "reasonable" value or adopt a so-called "best fitting ellipsoid." Hipkin (2002) has argued for the first approach: "To me it seems inevitable that, in the near future, we shall adopt a vertical reference system based on adopting a gravity model and one that incorporates $W = W_0 \equiv U_0$ to define its datum," with the justification that, "Nowadays, when observations are much more precise, their differences [between mean sea surface heights at various measuring stations] are distinguishable and present practice leads to confusion. It is now essential that we no longer associate mean sea level with any aspect of defining the geoid." In fact, previous gravimetric geoid models were computed by choosing to model a specific $W = W_0$ surface (Smith and Roman 2001). Defining $W_0 \equiv U_0$ is unnecessary because it is computable as the zero-order geoid undulation (Dru Smith 2006, personal communication). Other researchers have explored the second alternative by using the altimetry and GPS+leveling methods mentioned above. However, different level surfaces fill the needs of different user groups better than others. Moreover, defining a single potential value for all time is probably inadequate because mean sea level is constantly changing due to, for example, the changing amount of water in the oceans, plate tectonics changing the shape and volume of the ocean basins and the continents, and "thermal expansion of the oceans changing ocean density resulting in changing sea levels with little corresponding displacement of the equipotential surface" (Hipkin 2002). The geoid is constantly evolving, which leads to the need for episodic datum releases, as is done in the United States with mean sea level. A global vertical datum will only be adopted if it meets the needs of those who use it.

10.3 U.S. National Vertical Datums

The first leveling route in the United States considered to be of geodetic quality was established in 1856-57 under the direction of G. B. Vose of the U.S. Coast Survey, predecessor of the U.S. Coast and Geodetic Survey and, later, the National Ocean Service. The leveling survey was needed to support tide and current studies in the New York Bay and Hudson River areas. The first leveling line officially designated as "geodesic leveling" by the Coast and Geodetic Survey followed an arc of triangulation along the 39th parallel. This 1887 survey began at bench mark A in Hagerstown, Maryland.

By 1900, the vertical control network had grown to 21 095 km of geodetic leveling. A reference surface was determined in 1900 by holding elevations referenced to local mean sea level (LMSL) fixed at five tide stations. Data from two other tide stations indirectly influenced the determination of the reference surface. Subsequent readjustments of the leveling network were performed by the Coast and Geodetic Survey in 1903, 1907, and 1912 (Berry 1976).

10.3.1 National Geodetic Vertical Datum of 1929 (NGVD 29)

The next general adjustment of the vertical control network, called the Sea Level Datum of 1929 and later renamed to the National Geodetic Vertical Datum of 1929 (NGVD 29), was accomplished in 1929. By then, the international nature of geodetic networks was well understood, and Canada provided data for its First-Order vertical network to combine with the U.S. network. The two networks were connected at 24 locations through vertical control points (bench marks) from Maine/New Brunswick to Washington/British Columbia. Although Canada did not adopt the Sea Level Datum of 1929, Canadian-U.S. cooperation in the general adjustment greatly strengthened the 1929 network. Table 10.2 lists the kilometers of leveling involved in the readjustments and the number of tide stations used to establish the datums (Berry 1976).

Sea Level Datum of 1929 was renamed National Geodetic Vertical Datum of 1929 in the 1970s to eliminate all reference to sea level in the title. This was a change in name only; the mathematical and physical definitions of the datum established in 1929 remained unchanged.

Table 10.2: Amounts of leveling and numbers of tide stations involved in U.S. vertical datum adjustments.

Year of Adjustment	Kilometers of leveling	Number of tide stations
1900	21 095	5
1903	31 789	8
1907	38 359	8
1912	46 468	9
1929	75 159 (U.S.)	21 (U.S.)
	31 565 (Canada)	5 (Canada)

10.3.2 North American Vertical Datum of 1988 (NAVD 88)

The most recent general adjustment of the U.S. vertical control network, which is known as the North American Vertical Datum of 1988 (NAVD 88), was completed in June 1991 (Zilkoski et al. 1992). Approximately 625 000 km of leveling have been added to the National Spatial Reference System (NSRS) since NGVD 29 was created. In the intervening years, discussions were held periodically to determine the proper time for the inevitable new general adjustment. In the early 1970s, the National Geodetic Survey conducted an extensive inventory of the vertical control network. The search identified thousands of bench marks that had been destroyed, due primarily to post-World War II highway construction. Many existing bench marks were affected by crustal motion associated with earthquake activity, postglacial rebound (uplift), and subsidence resulting from the withdrawal of underground liquids.

An important feature of the NAVD 88 program was the releveling of much of the First-Order NGS vertical control network in the United States. The dynamic nature of the network requires a framework of newly observed height differences to obtain realistic, contemporary height values from the readjustment. To accomplish this, NGS identified 81 500 km (50 600 miles) for releveling. Replacement of disturbed and destroyed monuments preceded the actual leveling. This effort also included the establishment of stable "deep rod" bench marks, which are now providing reference points for new GPS-derived orthometric height projects as well as for traditional leveling projects. The general adjustment of NAVD 88 consisted of 709,000 unknowns (approximately 505 000 permanently monumented bench marks and 204 000 temporary bench marks) and approximately 1.2 million observations.

Analyses indicate that the overall differences for the conterminous United States between orthometric heights referred to NAVD 88 and NGVD 29 range from +40 to +150 cm. In Alaska the differences range from approximately +94 to +240 cm. However, in most stable areas, relative height changes between adjacent bench marks appear to be less than 1 cm. In many areas, a single bias factor, describing the difference between NGVD 29 and NAVD 88, can be estimated and used for most mapping applications (NGS has developed a program called VERTCON to convert from NGVD 29 to NAVD 88 to support mapping applications). The overall differences between dynamic heights referred to International Great Lakes Datum of 1985 (IGLD 85) and IGLD 55 range from 1 to 37 cm.

10.3.3 International Great Lakes Datum of 1985 (IGLD 85)

For the general adjustment of NAVD 88 and the International Great Lakes Datum of 1985 (IGLD 85), a minimum constraint adjustment of Canadian-Mexican-U.S. leveling observations was performed. The height of the primary tidal bench mark at Father Point/Rimouski, Quebec, Canada (also used in the NGVD 1929 general adjustment), was held fixed as the constraint. Therefore, IGLD 85 and NAVD 88 are one and the same. Father Point/Rimouski is an IGLD water-level station located at the mouth of the St. Lawrence River and is the reference station used for IGLD 85. This constraint satisfied the requirements of shifting the datum vertically to minimize the impact of NAVD 88 on U.S. Geological Survey (USGS) mapping products, and it provides the datum point desired by the IGLD Coordinating Committee for IGLD 85. The only difference between IGLD 85 and NAVD 88 is that IGLD 85 bench mark values are given in dynamic height units, and NAVD 88 values are given in Helmert orthometric height units. Geopotential numbers for individual bench marks are the same in both systems.

10.3.4 NAVD 88 and IGLD 85

Neither the NAVD 88 nor the IGLD 85 attempt to define the geoid or to realize some level surface which was thought to be the geoid. Instead, they are based upon a level surface that exists near the geoid but at some small, unknown distance from it. This level surface is situated such that shore locations with a height of zero in this reference frame will generally be near the surface of the ocean. IGLD 85 had a design goal that its heights be referenced to the water level gauge at the mouth of the St. Lawrence River. NAVD 88 had a design goal that it minimize recompilation of the USGS topographic map series, which was referred to NGVD 29. The station at Father Point/Rimouski met both requirements. NAVD 88 was realized using Helmert orthometric heights, whereas IGLD 85 employs dynamic heights. According to CCGLBHHD,

> Two systems, orthometric and dynamic heights, are relevant to the establishment of IGLD (1985) and NAVD (1988). The geopotential numbers for individual bench marks are the same in both height systems. The requirement in the Great Lakes basin to provide an accurate measurement of potential hydraulic head is the primary reason for adopting dynamic heights. It should be noted that dynamic heights are basically geopotential numbers scaled by a constant of 980.6199 gals, normal gravity at sea level at 45 degrees latitude. Therefore, dynamic heights are also an estimate of the hydraulic head.

> ...IGLD 85 and NAVD 88 are now one and the same...The only difference between IGLD 85 and NAVD 88 is that IGLD 85 bench mark values are given in dynamic height units, and NAVD 88 values are given in Helmert orthometric height units. The geopotential numbers of bench marks are the same in both systems.

Because the United States covers a large area north to south with a considerable variety of topographic features, dynamic heights would not be entirely acceptable, because the dynamic corrections in the interior of the country would often be unacceptably large. The United States is committed now and for the future to orthometric heights, which in turn implies a commitment to geoid determination.

10.3.5 Tidal datums

A vertical datum is called a tidal datum when it is defined by a certain phase of the tide. Tidal datums are local datums and are referenced to nearby monuments. Since a tidal datum is defined by a certain phase of the tide, there are many different types of tidal datums (Meyer et al. 2006a; Marmer 1951; Gill and Schultz 2001; Zilkoski 2001). The principal tidal datums typically used by federal, state, and local government agencies are Mean Higher High Water (MHHW), Mean High Water (MHW), Mean Sea Level (MSL), Mean Low Water (MLW), and Mean Lower Low Water (MLLW).

Principal tidal datums in the United States are based on the average of observations over a 19-year period, e.g., 1988-2001. A specific 19-year Metonic cycle is denoted as a National Tidal Datum Epoch (NTDE). CO-OPS publishes the official United States local mean sea level values as defined by observations at the 175-station National Water Level Observation Network (NWLON).

Principal tidal datums are defined as follows:

Mean Higher High Water MHHW is defined as the arithmetic mean of the higher high water heights of the tide observed over a specific 19-year Metonic cycle denoted as the NTDE. Only the higher high water of each pair of high waters of a tidal day is included in the mean. For stations with shorter series, a comparison of simultaneous observations is made with a primary control tide station in order to derive the equivalent of the 19-year value (Marmer 1951).

Mean High Water MHW is defined as the arithmetic mean of the high water heights observed over a specific 19-year Metonic cycle. For stations with shorter series, a computation of simultaneous observations is made with a primary control station in order to derive the equivalent of a 19-year value (Marmer 1951).

Mean Sea Level MSL is defined as the arithmetic mean of hourly heights observed over a specific 19-year Metonic cycle. Shorter series are specified in the name, like monthly mean sea level or yearly mean sea level (e.g., Marmer 1951; Hicks 1985).

Mean Low Water MLW is defined as the arithmetic mean of the low water heights observed over a specific 19-year Metonic cycle. For stations with shorter series, a comparison of simultaneous observations is made with a primary control tide station in order to derive the equivalent of a 19-year value (Marmer 1951).

Mean Lower Low Water MLLW is defined as the arithmetic mean of the lower low water heights of the tide observed over a specific 19-year Metonic cycle. Only the lower low water of each pair of low waters of a tidal day is included in the mean. For stations with shorter series, a comparison of simultaneous observations is made with a primary control tide station in order to derive the equivalent of a 19-year value (Marmer 1951).

Other tidal values typically computed include the Mean Tide Level (MTL), Diurnal Tide Level (DTL), Mean Range (Mn), Diurnal High Water Inequality (DHQ), Diurnal Low Water Inequality (DLQ), and Great Diurnal Range (Gt).

Mean Tide Level MTL is a tidal datum that is the average of Mean High Water and Mean Low Water.

10.4. YOUR TURN

Diurnal Tide Level DTL is a tidal datum that is the average of Mean Higher High Water and Mean Lower Low Water.

Mean Range Mn is the difference between Mean High Water and Mean Low Water.

Diurnal High Water Inequality DHQ is the difference between Mean Higher High Water and Mean High Water.

Diurnal Low Water Inequality DLQ is the difference between Mean Low Water and Mean Lower Low Water.

Great Diurnal Range Gt is the difference between Mean Higher High Water and Mean Lower Low Water.

Knowing which tidal datum the data are referenced to is important. Like geodetic vertical datums, local tidal datums are all different from one another, but the difference can be determined.

Local tidal datum 846 4336 (Middletown, Connecticut) is related to NAVD 88 as follows (see http://tidesandcurrents.noaa.gov/benchmarks/8464336.html):

MHHW	—	0.770 m
MHW	—	0.695 m
MSL	—	0.374 m
MTL	—	0.364 m
NAVD 88	—	0.186 m
MLW	—	0.033 m
MLLW	—	0.000 m

The primary bench mark (designation 846 4336 E) has no NGS permanent identifier (PID), and therefore, its NAVD 88 elevation is not known. However, two associated bench marks do have NAVD 88 First-Order elevations: MIDDLETOWN 1935 PID LX0583 has H = 15.233 m and Y 14 1935 PID LX0580 has H = 10.049 m. According to the Bench Mark Elevation Information given in the table, MIDDLETOWN 1935 is 15.414 m above MLLW and Y 14 1935 is 10.229 m above MLLW.

Figure 10.4 illustrates the geometry. The blue numbers are heights in meters above MLLW at the tide gauge. The numbers in black come from the Bench Mark Elevation Information table, which specifies offsets in meters from the bench marks to MLLW at this tidal station. The green numbers are NAVD 88 Helmert orthometric heights from NGS data sheets. NAVD 88 elevations can be deduced for the other tidal bench marks at this station by subtracting 0.186 m from their offset. The offset for 846 4336 E is 6.025 m, so its NAVD 88 height is 5.845 m (shown in parentheses).

NAVD 88 and the tide gauge disagree about the location of the NAVD 88 level surface by 6 mm, because the datums are independent. They can be expected to closely resemble one another but not to agree exactly.

10.4 Your Turn

Problem 10.1. Use GEOID09 to determine a geoid height for KERR (see problem 8.2 on page 168). Use Eq. 10.7 to determine KERR's orthometric height. Then, rework problem 8.2 on page 168

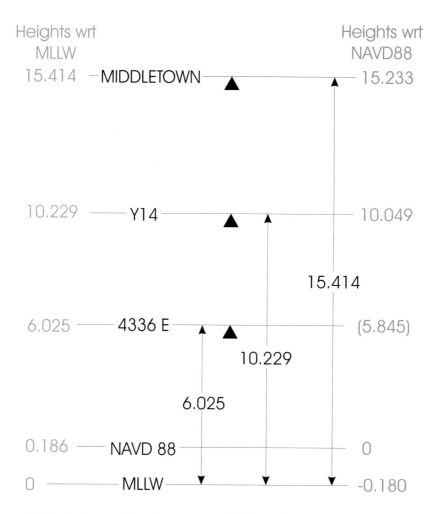

Figure 10.4: Relationships between a tidal bench mark and a tidal datum.

using Eq. 6.18 instead of the ellipsoid height reduction equations (Eq. 6.15 and 6.16). How far apart are your coordinates determined using orthometric heights and the simplified equation from your coordinates determined using ellipsoid heights and the fully rigorous equations?

Chapter 11

Tides

Stability over time is a desirable quality for a datum – this is a concept well understood by surveyors. With GNSS technology, orthometric heights can be determined at an accuracy on the order of centimeters or better. If the datums to which the height systems are referred vary by this amount or more, these effects must be taken into account and removed. We will now look at whether the geoid is stable over time, and if not, how quickly it changes and by how much.

An investigation into the variability of the geoid is equivalent to an investigation into the variability of the Earth's gravity potential field; it is a subject in the field of geodynamics. Changes in the Earth's gravity are caused by changes in (i) the Earth's diurnal rotation that produces the centrifugal force component of gravity, (ii) the Earth's mass and its distribution, or (iii) the spatial arrangement of objects massive enough and near enough that their gravitational fields have discernible effects on the geoid.

The Earth's diurnal rotation is not constant in velocity or direction. The length of the day is decreasing by about two milliseconds per century, and there are seasonal variations (with periods on the order of a month) on the same order (Vaníček and Krakiwsky 1996, p. 68). Consequently, the Earth's centrifugal force is likewise diminishing and variable. However, these variations are far too small (on the order of 10^{-12} radians s^{-1}) to change the Earth's centrifugal force at a discernible level in a time frame shorter than the geologic.

Precession and nutation change the direction of the Earth's rotational axis. There are additional, smaller perturbations as well. The motion of the Earth's rotational axis in the celestial reference frame affects astronomic and satellite observations but not gravity because, although the direction of the centrifugal force vector is changing, this change is brought about by a motion of the Earth itself, so the relative change is zero. However, actual movement of the rotational axis relative to the Earth's crust itself (known as polar motion or polar wobble) does affect gravity, because the direction of the centrifugal force vector in this case is changing relative to the Earth's crust. These small changes are only on the order of a few nanogals, well below the noise level of most gravity measurements.

The Earth's mass can increase or decrease, and it can be redistributed. The Earth gains mass almost continuously due to a stream of space debris entering the atmosphere and, occasionally, striking the Earth's surface. Similarly, the Earth is constantly losing mass as gaseous molecules too light to be bound by gravity drift off into space (e.g., helium gas). Neither the addition nor the removal of mass changes the Earth's gravity field enough to be of concern in this context.

The Earth's mass is redistributed in various ways including post-glacial rebound, melting ice caps and glaciers, the Earth's fluid outer core, the oceans (Cazenave and Nerem 2002), and earthquakes. For example, earthquakes can be caused by the motion of tectonic plates along their margins, and this motion causes a change in the Earth's shape. Earthquakes can cause a measurable change in the Earth's rotational velocity, and thus its gravity, by changing one of its moments of inertia (Chao and Gross 1987; Smylie and Manshina 1971; Soldati and Spada 1999). The Sumatra, Indonesia, earthquake of December 26, 2004, was such an event. It decreased the length of day by 2.68 microseconds, shifted the "mean North Pole" about 2.5 cm in the direction of 145 degrees east longitude, and decreased the Earth's flattening by about one part in 10 billion (Buis 2005). The uplift of plates due to tectonic or post-glacial activities affects ellipsoidal heights, as well as having a smaller gravity-based effect which changes the geoid.

People who have been at an ocean shore for half a day or more have had the opportunity to watch the ocean advance inland and then retreat back out to sea. This motion is caused primarily by the gravitational attraction of the Moon and, to a lesser degree, the Sun. Therefore, the definition of tide found in the Geodetic Glossary (NGS 2009) may be somewhat surprising:

> **Tide** (1) Periodic changes in the shape of the Earth, other planets or their moons that relate to the positions of the Sun, Moon, and other members of the solar system.

Note that this definition is not about the oceans, per se. Instead, it speaks of, among other things, a change of the shape of the Earth itself, the **earth tide** or **body tide**. It is common knowledge that the Moon moves the oceans; it deforms them to set them in motion. But, what is probably not so well known is that the Earth's core, mantle, and crust have their shape deformed in a manner similar to the deformation of the oceans. The Glossary continues:

> In particular, (2) those changes in the size and shape of a body that are caused by movement through the gravitational field of another body. The word is most frequently used to refer to changes in size and shape of the Earth in response to the gravitational attractions of the other members of the solar system, in particular, the Moon and Sun. In such cases, three different tides are usually distinguished: the atmospheric tide, which acts on the gaseous envelope of the Earth; the Earth tide, which acts on the solid Earth; and the ocean tide (usually simply called "the tide"), which acts on the hydrosphere.

The effects of the tides are numerous and complicated, so perhaps the first question to consider is whether the tides cause enough of an effect to be of concern. Is the Earth tide large enough to affect the geoid in any practical way? There are two high and low earth tides each day, with the highest being on the order of a 50-cm displacement (Moritz 1980, p. 477)! Clearly, tides must be taken into consideration.

Tides on the Earth arise due to the influences from all celestial bodies. The Sun and the Moon produce the largest effects by far, but the other planets have discernible affects, albeit too small to impact GNSS positioning (Wilhelm and Wenzel 1997, p. 11). All celestial bodies create tides in the same way, the only difference being the details of how these manifest themselves. Therefore, we will consider the effects created by the Moon, with the understanding that they apply to any celestial body, with the appropriate change of mass and distance variables.

11.1 Tidal Gravitational Attraction and Potential

According to Newton, force gives rise to motion by accelerating mass. The gravitational force of the Moon on the Earth itself is found using Eq. 9.1 on page 174:

$$\mathbf{F}_m = -\frac{GMm\hat{\mathbf{r}}}{|\mathbf{r}|^2}, \tag{11.1}$$

where $\mathbf{r}$ is a vector from the Moon's center to the Earth's center (note the negative sign in Eq. 11.1 reversing the direction of the vector so that the force is directed from the Earth's center toward the Moon's center); M and m are the masses of the Earth and the Moon, respectively; and $\mathbf{F}_m$ is the gravitational force vector produced by the Moon on the Earth.

The gravitational force of the Earth exerted on the Moon can be found simply by defining $\mathbf{r}$ to have the opposite direction, so the magnitudes of the two forces are equal. The gravitational attraction of the Earth on the Moon causes the Moon to orbit the Earth rather than to move off into space. However, Eq. 11.1 also means that the Earth is orbiting the Moon, but this motion is much less obvious due to the difference in masses of the two bodies. If we take 5.9742×10^{27} g to be the Earth's mass, 7.38×10^{25} g to be the Moon's mass, and 3.84×10^8 m to be their mean separation, then the barycenter of the Earth–Moon system can be found to be at a point on a line connecting their two centers approximately 4.69×10^6 m from the Earth's center. This point is inside the Earth, being about 73.5 percent of the length of the GRS 80 semimajor axis.

Points on different radii on a rigid rotating body move in different directions and at different instantaneous linear velocities (see Fig. 11.1.a). However, a rigid body can rotate around only one axis at any moment in time. The Earth rotates around its axis, but it does not rotate about the Earth–Moon barycenter; instead, it *revolves* around it. **Revolution** is the motion that brings an object back around to a place it has been before. A merry-go-round rotates around its shaft but riders on a merry-go-round revolve around the shaft. For the Earth–Moon system, one can envision both the Earth and the Moon being on a large merry-go-round, with the Earth placed so that the merry-go-round's axis is underneath the Earth and the Moon set on the merry-go-round's edge. Passengers on the merry-go-round's edge experience a constant centrifugal force directed away from the axis of rotation. Thus, the orbital rotation of the Earth around the Earth–Moon barycenter gives rise to a constant acceleration that is always directed opposite to the line connecting the Earth's center to the Moon's. In particular, everywhere and everything on and in the Earth is accelerating away from the Moon as if the Earth were moving in a straight line along the instantaneous axis between them (Fig. 11.1.b). This acceleration gives rise to a component of observable gravity that is at most 3.4 percent of the total acceleration (Vaníček and Krakiwsky 1996, p. 125).

The complete motion of the Earth–Moon–Sun system could be modeled as

- the Earth centered on its own merry-go-round to model its diurnal rotation,

- the rotating Earth set atop a merry-go-round at the Earth–Moon system so that the Earth's axis of rotation is offset from the Earth–Moon barycenter as described above, and

- the Earth–Moon merry-go-round set atop another merry-go-round so as to model the revolution around the solar system's barycenter near the Sun's center of mass.

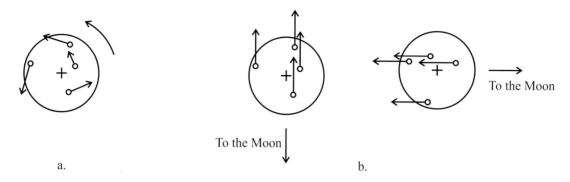

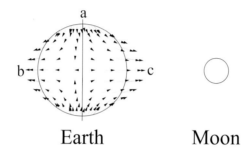

Figure 11.1: (a) Instantaneous velocity vectors of four places on the Earth. The acceleration vectors (not shown) would be perpendicular to the velocity vectors directed radially toward the rotation axis. The magnitudes and directions of these velocities are functions of the distances and directions to the rotation axis (+). (b) Acceleration vectors of the same places at two different times of the month. The acceleration magnitude is constant, and its direction is always away from the Moon.

Figure 11.2: Arrows indicate force vectors that are the combination of the Moon's attraction and the Earth's orbital acceleration around the Earth–Moon barycenter (based on Bearman 1999, p. 54–56 and Vaníček and Krakiwsky 1996, p. 124).

The Moon's gravitational attraction gives rise to a force at any particular place on the Earth that is directed (approximately[1]) along the line from the point of interest to the Moon's center. In contrast, the orbital acceleration experienced at that place is always parallel to the line connecting the Earth–Moon centers, so these forces are not generally parallel to each other. Furthermore, places on the side of the Earth facing away from the Moon experience a smaller attraction than places on the side facing towards the Moon due to being closer to the Moon, giving rise to the asymmetry shown in Fig. 11.2. Each of the vectors in Fig. 11.2 represents the force resulting from the combination of the orbital acceleration and the Moon's attraction at the place indicated by the tail of the vector.

Figure 11.3 shows the vector addition details of three points of interest from Fig. 11.2. Orange vectors are the Moon's attraction; their nonparallelism with the orbital acceleration vectors, shown in blue, is greatly exaggerated in the figure. The vector result of the addition of these two vectors is

---

[1] The Moon is too close to the Earth for this to be exact. The actual direction of the vector would be determined by triple integrating over the Moon's mass, and approximately (because the Moon is not a perfectly homogeneous sphere) ending up pointing at the Moon's center of mass.

11.1. TIDAL GRAVITATIONAL ATTRACTION AND POTENTIAL

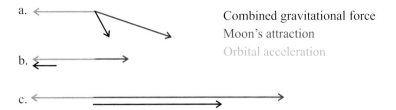

Figure 11.3: Combined gravitational force at the North Pole (a), at a point on Earth furthest from the Moon (b), and at a point on Earth closest to the Moon (c).

shown in black. Point **a** is located at the top of the circle in Fig. 11.2. The Moon's attraction is the least parallel with the orbital acceleration at this place and its antipodal counterpart. Given the roughly equal magnitudes of the orbital acceleration and Moon's attraction forces, their components in the direction of the Moon largely cancel at **a**, leaving a small vector oriented sharply toward the Earth's middle. Point **b** is furthest from the Moon. The Moon's attraction is parallel to the orbital acceleration at this point, but opposite in direction. The orbital acceleration is moderately stronger than the Moon's attraction here, creating the force primarily responsible for the lower high tide of the day. Point **c** is located closest to the Moon. The Moon's attraction is considerably stronger here than the orbital acceleration, creating the force that is primarily responsible for the higher tide of the day (Vaníček and Krakiwsky 1996, p. 124, Bearman 1999, p. 52–61).

The magnitude and direction of the Moon's attraction are periodic due to the nature of the Moon's orbit around the Earth. The liquid nature of the oceans allows dramatically more complexity in their response to gravitational attraction and, consequently, its modeling is likewise more complex. The situation is complicated but can be made tractable by accounting for individual **tidal constituents**, which are sinusoids with a particular amplitude, frequency, and phase that arises due to a particular phenomenon. It is possible to decompose the Moon's attraction into individual constituents (Boon 2004). Some of the prominent tidal constituents are caused by

- the inclination of the Moon's orbital plane with respect to the ecliptic giving rise to the lunar declination (tropic-equatorial) cycle,

- the Sun's attraction giving rise to the spring–neap cycle,

- the eccentricity of the Moon's orbit giving rise to the perigean–apogean cycle, and

- the precession of the lunar nodes giving rise to the Metonic cycle.

Simple ocean tide models include as few as six constituents; complicated models can incorporate more than 100 (Wilhelm and Wenzel 1997). These models produce tidal predictions such as those shown in Fig. 11.4, which use constituents given by Boon (2004, p. 97–102) and clearly show higher high water, lower high water, higher lower water, and lower low water, as well as many longer-period variations.

Up to this point we have been concerned with gravity force. We now consider how tides affect gravity potential because, after all, the geoid (an equipotential surface) is a principal datum of interest. The gravitational potential field created by the Moon at some point of interest can be expressed as an infinite series of which only the second term is important for tides. This second

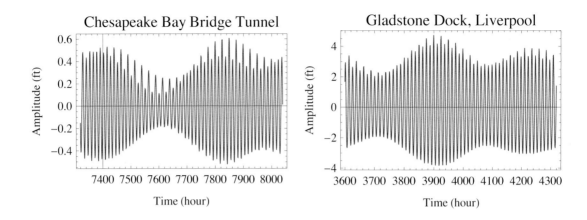

Figure 11.4: Two simulations of tide cycles illustrating the variety of possible effects.

term W_2 takes the following form (Vaníček 1980, p. 5, Eq. 12):

$$W_2 \approx D \left[\overbrace{\cos^2 \psi \cos^2 \delta \cos 2t}^{\text{sectorial}} + \overbrace{\sin 2\psi \sin 2\delta \cos t}^{\text{tesseral}} + \overbrace{3(\sin^2 \psi - 1/3)(\sin^2 \delta - 1/3)}^{\text{zonal}} \right] \quad (11.2)$$

where D is Doodson's constant (Doodson 1922), ψ is geocentric latitude, δ is the declination of the Moon (Meeus 1998), and t is the Moon's hour angle (Meeus 1998). Doodson's constant is given by Vaníček (1980, p. 4, Eq. 7) as

$$D = \frac{3}{4} Gm \frac{R^2}{r_m^3} \quad (11.3)$$

where G is the universal gravitation constant, m is the mass of the Moon, R is the mean (equivoluminous) radius of the Earth, and r_m is the mean distance to the Moon. D has a value of approximately 2.6277×10^7 cm mgal.

Equation 11.2 consists of three terms within the brackets. The first term contains sectorial constituents, the second term contains tesseral constituents, and the third term contains zonal constituents. These three components are shown in Fig. 11.5 in panels a, b, and c, and their combination in Fig. 11.5.d. Sectorial constituents vary in longitude (time), like the sectors of an orange, and give rise to the two daily tides. Figure 11.5.b shows the zonal constituents. The red and blue lines are as in Fig. 11.5.a, but the equatorial green line is entirely inside the potential surface. The zonal constituent gives rise to latitudinal tides because it is a function of latitude. The zonal constituents do not vary in time and give rise to so-called permanent tides. Tesseral constituents possess both latitude and longitude components and give rise to patterns resembling the tessellation of a checkerboard. The tesseral constituent gives rise to both longitudinal and latitudinal tides, producing a somewhat distorted result, which is highly exaggerated in the figure for clarity. The tesseral constituent accounts for the Moon's orbital plane being inclined by about 5 degrees from the plane of the ecliptic.

11.1. TIDAL GRAVITATIONAL ATTRACTION AND POTENTIAL 217

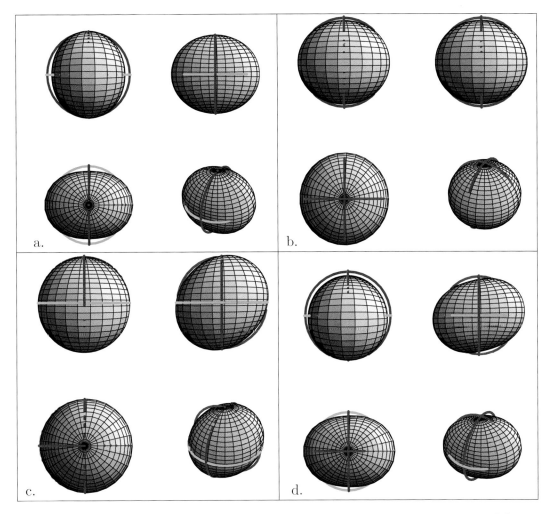

Figure 11.5: Sectorial (a), zonal (b), and tesseral (c) constituents of tidal potential. (d) Total tidal potential is the combination of the sectorial, zonal, and tesseral constituents. The equator (green), the prime meridian and international date line (red), and the 90° and 270° meridians (blue) are indicated.

11.2 Ocean Tides and Body Tides

Ocean tides affect the geoid by redistributing the mass of the oceans, which creates a discernible change in the geoid. The first clear evidence of body tides came from the measurement of ocean tides, which were consistently about two-thirds as high as Newton's physics predicted. The missing one-third was due to deformation of the Earth itself, moving with the oceans (Melchior 1974). The weight of the water deforms the Earth below it, in addition to the tidal potential also deforming the Earth (Vaníček 1980, p. 9–12). The tides of the solid Earth behave in the same manner as the ocean tides, but in a simpler manner because the Earth deforms like an elastic solid at the frequencies of tides, rather than with all the freedom of a liquid, like the oceans.

It is remarkable that the effect of the Moon's potential field upon the Earth can be described with such high accuracy by such a relatively simple equation as Eq. 11.2, especially in comparison with the effort necessary to determine the geoid! The simplicity of Eq. 11.2 is due to (i) the Moon being far enough away to be treated as a point mass, and (ii) the motion of the Moon being very accurately described by celestial mechanics. Therefore, no gravity observations are needed to determine the potential from the Moon; it all falls out of the mathematics.

The parameters that describe the response of the Earth's shape and gravitational potential field to tidal forces are called **Love** and **Shida** numbers, which are empirically derived. They are used in equations similar to Eq. 11.2 and sufficiently capture the deformation of the Earth so that tidal effects may be removed from geoid models, gravity observations, GNSS observations, and other geodetic quantities (Vaníček 1980). However, permanent tides (those portions of the tidal equations which describe the nontime-varying deformations) are not completely determinable empirically. The permanent tide has two components: (i) the permanent deformation of Earth's geopotential field due to the existence of the permanent (nonzero time-averaged) Sun and Moon, and (ii) the permanent deformation of Earth's geopotential field due to the existence of the permanent deformation of Earth's crust (which, in turn, is due to the existence of the permanent Sun and Moon).

The first part (called the "direct" component of the permanent Earth tide) is computable empirically, as it deals solely with the Sun's and Moon's masses affecting the Earth's geopotential field. The second part is not computable empirically. This is because the permanent deformation of the Earth's crust cannot be directly observed. The Earth's crust perpetually (permanently) exhibits a deformation due to the permanent existence of the Sun and Moon. Because we cannot observe how the crust would react without a permanent Sun and Moon, we cannot determine empirically how much permanent deformation actually exists, and thus cannot compute the effect of this permanent crustal deformation on the Earth's geopotential.

11.3 *Your Turn*

Problem 11.1. Imagine our solar system was exactly as it is but without the Moon. Would there still be tides on the Earth?

Problem 11.2. An object with mass m (kg) is moving in a circular path with radius r (m) at a constant angular velocity ω (s^{-1}, radians per second). Let $\mathbf{r}$ denote the object's position vector, directed outwards, with respect to the circle's center, so $|\mathbf{r}| = r$. The object generates a centrifugal force $\mathbf{f}_c = m\,\omega^2/\mathbf{r}$ in newtons. Use reduced latitude to graph (or tabulate) the magnitude of the centrifugal force on a unit mass at the surface of the GRS 80 reference ellipsoid from the equator to the North Pole. You can find the Earth's angular velocity in section 9.4 on page 185.

Problem 11.3. The universal gravitational constant $G = 6.67259 \times 10^{-11}$ m^3 s$^{-2}kg^{-1}$. The Moon's mass is 7.38×10^{23} kg. Graph (or tabulate) the magnitude of the gravitational force $\mathbf{f}_m$ exerted by the Moon on a unit mass on the closer surface of the GRS 80 reference ellipsoid from the equator to the North Pole if the Moon is in the Earth's equatorial plane and at an Earth-centroid to Moon-centroid distance of 384 405 km.

Problem 11.4. Graph (or tabulate) the magnitude of the combined force $\mathbf{f}_m + \mathbf{f}_c$ on the closer side of the GRS 80 reference ellipsoid from the equator to the North Pole. This computation requires you to compute the force vector first, and then determine its magnitude because $\mathbf{f}_m$ is not parallel to $\mathbf{f}_c$.

Problem 11.5. Graph (or tabulate) the magnitude of the combined force $\mathbf{f}_m + \mathbf{f}_c$ on the farther side of the GRS 80 reference ellipsoid from the equator to the North Pole.

Part IV

Appendixes

Appendix A

Vector Algebra and Linear Algebra

This appendix is a brief overview of vector algebra and linear algebra. There are many textbooks written on these subjects (Davis and Snider 1979; Boehm and Prautzsch 1994; Strang and Borre 1997).

A.1 Definitions and Arithmetic Operations

A point is a fundamental geometric object. Euclid's *Elements* begin with the definition of point. Euclid defined a point as "that which has no part" (Densmore and Heath 2000). In modern parlance, a point is a zero dimension object. The point's location can be represented by a position in one of any number of coordinate systems.

Points can be subtracted but they cannot be added. A **vector** is a geometric object that exactly characterizes the directed separation between two points. The separation of two scalars is given arithmetically by their difference: 10 is separated from 8 by 10 - 8 = 2. A vector is defined as the difference of two points and computed by their arithmetic difference. The vector **V** from **B** to **A** is **V** = **A** − **B** (the *to* position minus the *from* position).

Example A.1. Position **A** = (10, 30), and **B** = (-20, 45). Compute the vector from **B** to **A**.
Solution: The vector from **B** to **A** is **V** = **A** − **B** = $(10, 30) - (-20, 45) = (30, -15)$.
□

The vector from **B** to **A** is arithmetically the negative of the vector from **A** to **B**: **A** − **B** = −(**B** − **A**). The direction of a vector can be reversed by changing the sign of all its coordinates, written as -**A** and called **unary negation**.

Adding two vectors produces another vector. Adding a vector to a point produces another point. This is the logical inverse of the definition of a vector, being the difference of two points. Symbolically, if **V** = **B** - **A**, then **B** = **A** + **V**.

The **magnitude** of a vector is its length and is written |**V**|. The length of a vector is computed using the Pythagorean theorem (Eq. 3.6 on page 30).

Example A.2. Compute the magnitude of **V** from Example A.1.
Solution: The magnitude of **V** from Example A.1 is $\sqrt{30^2 + (-15)^2} = 33.541$.
□

Vector subtraction is defined as the addition of two vectors, one of which has had unary negation applied to it. In effect, this is adding a negative: **A** - **B** = **A** + (-**B**).

Example A.3. Suppose the WGS 84(G1150) XYZ position of GPS SV1 is (6 105 719.837, −18 791 473.420, 17 762 042.080), and GPS SV2 is at (6 970 183.991, 4 355 454.353, -25 255 027.900). How far apart are they?

Solution: If **V** is the vector from SV2 to SV1, then

$$
\begin{aligned}
|\mathbf{V}| &= |SV1 + (-SV2)| \\
&= |(6\,105\,719.837, -18\,791\,473.420, 17\,762\,042.080) + (-6\,970\,183.991, -4\,355\,454.353, 25\,255\,027.900)| \\
&= |(-864\,464.154, -23\,146\,927.780, 43\,017\,069.990)| \\
&= \sqrt{-864\,464.154^2 + (-23\,146\,927.780)^2 + 43\,017\,069.990^2} \\
&= 48\,856\,891.780
\end{aligned}
$$

□

A **unit vector** is any vector for which $|\mathbf{V}| = 1$. Unit vectors are denoted as $\hat{\mathbf{V}}$. So, $\hat{\mathbf{V}} = \mathbf{V}/|\mathbf{V}|$.

The **vector inner product** of **A** and **B** is written **A.B**, and it produces a scalar:

$$\mathbf{A.B} = \cos\theta |\mathbf{A}||\mathbf{B}|$$

or

$$\cos\theta = \frac{\mathbf{A.B}}{|\mathbf{A}||\mathbf{B}|} \tag{A.1}$$

If **A** and **B** are unit vectors

$$\cos\theta = \mathbf{A.B} \tag{A.2}$$

and $\theta = \arccos(\mathbf{A.B})$. If **A** and **B** both have n coordinates, then

$$\mathbf{A.B} = \sum_{i=1}^{n} a_i b_i \tag{A.3}$$

Example A.4. What is the inner product of the vectors in Example A.3?
Solution: From Eq. A.3:

$$
\begin{aligned}
\mathbf{SV1.SV2} &= \sum_{i=1}^{n} SV1_i SV2_i \\
&= 6\,105\,719.837 \cdot 6\,970\,183.991 + \\
&\quad (-18\,791\,473.420) \cdot 4\,355\,454.353 + \\
&\quad 17\,762\,042.080 \cdot (-25\,255\,027.900) \\
&= -487\,868\,282\,500\,000
\end{aligned}
$$

□

The magnitude squared of a vector is the inner product of the vector with itself: $|\mathbf{V}|^2 = \mathbf{V.V}$.

Example A.5. What is the angle between the vectors in Example A.3?
Solution: The angle is given by Eq. A.1, which depends on the magnitudes of the vectors: $|\mathbf{SV1}| = 26\,568\,579.700$ and $|\mathbf{SV2}| = 26\,558\,800.460$.

$$\cos\theta = \frac{\mathbf{SV1}.\mathbf{SV2}}{|\mathbf{SV1}||\mathbf{SV2}|}$$
$$= -487\,868\,282\,500\,000/(26\,568\,579.700 \cdot 26\,558\,800.460)$$
$$= -0.691\,394\,292$$

Thus, $\theta = \arccos(-0.691\,394\,292) = 133.741°$.
□

The vector outer product of vectors $\mathbf{A}$ and $\mathbf{B}$ is written $\mathbf{A} \times \mathbf{B}$, and it produces a vector perpendicular to the plane containing $\mathbf{A}$ and $\mathbf{B}$. Vector outer products (also called cross products) can only be performed on three-dimensional vectors. **Vector outer products** are computed by

$$\mathbf{A} \times \mathbf{B} = [a_2 b_3 - b_2 a_3,\ b_1 a_3 - a_1 b_3,\ a_1 b_2 - b_1 a_2] \tag{A.4}$$

Example A.6. Suppose $\mathbf{A} = [3, 5, 9]$ and $\mathbf{B} = $ [-1, -2, 6]. What is $\mathbf{A} \times \mathbf{B}$?
Solution: From Eq. A.4:

$$\mathbf{SV1}.\mathbf{SV2} = [a_2 b_3 - b_2 a_3,\ b_1 a_3 - a_1 b_3,\ a_1 b_2 - b_1 a_2]$$
$$= [5 \cdot 6 - (-2) \cdot 9,\ (-1) \cdot 9 - 3 \cdot 6,\ 3 \cdot (-2) - (-1) \cdot 5]$$
$$= [48, -27, -1]$$

□

The spatial angle between two vectors is related to the magnitude of the cross product by

$$|\mathbf{A} \times \mathbf{B}| = |\mathbf{A}| \cdot |\mathbf{B}| \sin\theta \tag{A.5}$$

Example A.7. What is the angle between the vectors in Example A.6?
Solution: From Eq. A.5:

$$|\mathbf{A} \times \mathbf{B}| = |\mathbf{A}| \cdot |\mathbf{B}| \sin\theta$$
$$\sin\theta = |\mathbf{A} \times \mathbf{B}|/(|\mathbf{A}| \cdot |\mathbf{B}|)$$
$$= \sqrt{3040}/(\sqrt{115} \cdot \sqrt{41})$$
$$= 0.802171$$

Thus, $\theta = \arcsin(0.802171) = 53.3379°$.
□

A.2 Vectors and Matrices

Vectors can be written as **column vectors**

$$\begin{bmatrix} x \\ y \\ z \end{bmatrix}$$

or as **row vectors**
$$[x, y, z]$$

The operation that takes them from one to the other is **transposition**, which is denoted by a capital T where an exponent would go. For example

$$[x, y, z]^T = \begin{bmatrix} x \\ y \\ z \end{bmatrix}$$

Vectors are column vectors by default. So, **v** is a column vector, and $\mathbf{v}^T$ is a row vector. Vectors are lists of scalars. A vector of vectors is called a **matrix** and might look like this:

$$\mathbf{M} = \begin{bmatrix} m_{11} & m_{12} & m_{13} \\ m_{21} & m_{22} & m_{23} \\ m_{31} & m_{32} & m_{33} \end{bmatrix}$$

The subscripts indicate an element's position in the matrix, with the row being the first number and column being the second. Thus, $m_{1,3}$ is the element in the first row and third column.

Matrices can have any number of rows and columns. The number of rows and the number of columns of a matrix constitute that matrix's **dimension**. For example, **M** has three rows and columns, so **M**'s dimension is 3×3. A matrix with two rows and three columns is 2×3. Matrices with equal numbers of rows and columns are called **square**. Column vectors can be thought of as $n \times 1$ matrices. Row vectors can be thought of as $1 \times n$ matrices.

Transposition operates on matrices just as it does for vectors; it transposes columns into rows. For **M** above, $\mathbf{M}^T$ is

$$\mathbf{M}^T = \begin{bmatrix} m_{11} & m_{21} & m_{31} \\ m_{12} & m_{22} & m_{32} \\ m_{13} & m_{23} & m_{33} \end{bmatrix}$$

Transposition reverses the order of the dimension of a matrix. For example, if **M** is 3×2, then $\mathbf{M}^T$ is 2×3.

The inner product of a matrix **M** and a column vector **v** is written as **M.v**. A matrix can be thought of as a vector of row vectors $\mathbf{M} = [\mathbf{m}_1^T, \mathbf{m}_2^T, \mathbf{m}_3^T]$, so the inner product $\mathbf{x} = \mathbf{M.v}$ is a vector whose elements are the inner product of each vector in the row matrix with **v**: $\mathbf{x}^T = [\mathbf{m}_1.\mathbf{v}, \mathbf{m}_2.\mathbf{v}, \mathbf{m}_3.\mathbf{v}]$. For example, if

$$\mathbf{M} = \begin{bmatrix} 1 & 5 \\ -3 & 2 \end{bmatrix}$$

and $\mathbf{v}^T = [-3, 4]$, then $\mathbf{M.v} = [\mathbf{m}_1.\mathbf{v}, \mathbf{m}_2.\mathbf{v}]^T = [1 \cdot (-3) + 5 \cdot 4, -3 \cdot (-3) + 2 \cdot 4]^T = [17, 16]^T$.

The inner product of two matrices **M** and **N** is written **M.N**. The number of columns for **M** must equal the number of rows for **N**. **M.N** is formed in the same way that matrices are multiplied by vectors: form the inner product of the ith row of **M** with the jth column of **N** as the i, j element of the product. If **M** is of dimension $i \times j$ and **N** is of dimension $j \times k$, then **M.N** is of dimension $i \times k$. The inner product of matrices is generally not commutative: $\mathbf{M.N} \neq \mathbf{N.M}$. In fact, unless both **M** and **N** are square and of the same dimension, one of those products cannot be formed.

A **zero matrix** has all zeros for elements. The product of a zero matrix with any other matrix, including vectors, is another zero matrix.

A **diagonal matrix** has nonzero elements along its diagonal and zeros elsewhere. If $\mathbf{D}$ is a diagonal matrix, then $d_{i,i} \neq 0$ and $d_{i,j} = 0$ for all $i \neq j$. An **identity matrix** is a diagonal matrix whose diagonal elements all equal 1. Identity matrices are usually denoted by $\mathbf{I}$. The product of an indentity matrix and any other matrix is the other matrix: $\mathbf{M.I} = \mathbf{M}$.

A matrix $\mathbf{M}^{-1}$ is the **inverse** of $\mathbf{M}$ if and only if $\mathbf{M.M}^{-1} = \mathbf{I}$. Inverse matrices are extremely important in linear algebra. Computing them is difficult, and not all matrices have inverses. However, inverse matrices allow us to solve linear algebra expressions such as the following: given $\mathbf{M}$ and $\mathbf{x}$ such that $\mathbf{M.v} = \mathbf{x}$, find $\mathbf{v}$. First, premultiply both sides of the equation by $\mathbf{M}^{-1}$, yielding $\mathbf{M}^{-1}.\mathbf{M.v} = \mathbf{M}^{-1}.\mathbf{x}$. Since $\mathbf{M.M}^{-1} = \mathbf{I}$, we have $\mathbf{I.v} = \mathbf{M}^{-1}.\mathbf{x}$, but $\mathbf{I.v} = \mathbf{v}$, so we end up with $\mathbf{v} = \mathbf{M}^{-1}.\mathbf{x}$. This simple matrix expression gives us the power to solve large systems of simultaneous linear equations very easily, provided we have a computer program to find matrix inverses.

A.3 Applications

Suppose $\mathbf{A}$ is an observation station, and $\mathbf{v}$ is a vector whose length and azimuth have been determined from total station observations. Then $\mathbf{B} = \mathbf{A} + \mathbf{v}$ is the forward problem of plane surveying. This vector notation is much more convenient and compact than the formulæ given in chapter 3.

Vector notation provides the products of inversing directly. The horizontal distance between two positions is the two-dimensional vector magnitude of their difference. The slope distance between two positions is the three-dimensional vector magnitude of their difference. The azimuth from one station $\mathbf{A}$ to another $\mathbf{B}$ comes from the inner product of the vector separation of the stations with a unit vector for north $[0, 1]$: $(\mathbf{B} - \mathbf{A}).[0, 1]$, which still needs to be disambiguated for quadrant, as usual.

Appendix B

Spherical Trigonometric Identities

The planar laws of sines and cosines are fundamental to plane surveying. These laws have spherical equivalents (Fig. 7.4 on page 139). Let **a** be the vector **OA**; likewise for **b** and **c**. Then, by the law of cosines, $\cos c = \mathbf{a}.\mathbf{b}$ because $|\mathbf{a}| = |\mathbf{b}| = 1$. Likewise, $\cos b = \mathbf{a}.\mathbf{c}$ and $\cos a = \mathbf{c}.\mathbf{b}$.

The spherical angle α is the dihedral angle formed between planes **AOB** and **AOC**. Similarly, β is the dihedral angle formed between planes **AOB** and **BOC**, and γ is the dihedral angle formed between planes **AOC** and **BOC**. These spherical angles can be determined by the inner product of the vectors normal to the planes defining the dihedral. The vector normal to **AOB** is $\mathbf{c}^\perp = \mathbf{OA} \times \mathbf{OB}$, the vector normal to **BOC** is $\mathbf{a}^\perp \equiv \mathbf{OB} \times \mathbf{OC}$, and the vector normal to **AOC** is $\mathbf{b}^\perp = \mathbf{OA} \times \mathbf{OC}$. For α, starting from the definition of the scalar product of the two vectors normal to its defining dihedral planes

$$
\begin{aligned}
\mathbf{c}^\perp.\mathbf{b}^\perp &= (\mathbf{a} \times \mathbf{b}).(\mathbf{a} \times \mathbf{c}) \\
&= (|\mathbf{a}||\mathbf{b}|\sin c)(|\mathbf{a}||\mathbf{c}|\sin b)\cos\alpha \\
&= \sin c \sin b \cos\alpha
\end{aligned}
\tag{B.1}
$$

Similarly

$$
\begin{aligned}
\mathbf{c}^\perp.\mathbf{a}^\perp &= \sin c \sin a \cos\beta \\
\mathbf{b}^\perp.\mathbf{a}^\perp &= \sin b \sin a \cos\gamma
\end{aligned}
$$

The cross product of the normal vectors is not readily obtainable, so we need to replace that expression with others in terms of the angles. In a vector algebra expression involving inner products and cross products, the inner and cross operations can be changed freely: $\mathbf{a}.\mathbf{b} \times \mathbf{c} = \mathbf{c}.\mathbf{a} \times \mathbf{b} = \mathbf{a} \times \mathbf{b}.\mathbf{c} = \mathbf{c} \times \mathbf{a}.\mathbf{b}$ (with the other combinations being different in sign). Since $\mathbf{c}^\perp = \mathbf{a} \times \mathbf{b}$, and $\mathbf{b}^\perp = \mathbf{a} \times \mathbf{c}$

$$
\begin{aligned}
\mathbf{c}^\perp.\mathbf{b}^\perp &= (\mathbf{a} \times \mathbf{b}).(\mathbf{a} \times \mathbf{c}) \\
&= \mathbf{a}.(\mathbf{b} \times (\mathbf{a} \times \mathbf{c})) \\
&= \mathbf{a}.((\mathbf{b}.\mathbf{c})\mathbf{a} - (\mathbf{b}.\mathbf{a})\mathbf{c}) \\
&= \mathbf{b}.\mathbf{c} - (\mathbf{b}.\mathbf{a})(\mathbf{a}.\mathbf{c}) \\
&= \cos a - \cos c \cos b
\end{aligned}
\tag{B.2}
$$

Setting Eq. B.1 equal to Eq. B.2, we get

$$\sin c \sin b \cos \alpha = \cos a - \cos c \cos b$$
$$\cos a = \cos b \cos c + \sin b \sin c \cos \alpha$$

which leads to

$$\cos b = \cos a \cos c + \sin a \sin c \cos \beta$$
$$\cos c = \cos a \cos b + \sin a \sin b \cos \gamma$$

These equations are called the **cosine rule for sides**. The analog to the law of sines can be written

$$\frac{\sin \alpha}{\sin a} = \frac{\sin \beta}{\sin b} = \frac{\sin \gamma}{\sin c}$$

The spherical version of the law of cosines has the form

$$\cos \alpha = -\cos \beta \cos \gamma + \sin \beta \sin \gamma \cos a$$
$$\cos \beta = -\cos \alpha \cos \gamma + \sin \alpha \sin \gamma \cos b$$
$$\cos \gamma = -\cos \alpha \cos \beta + \sin \alpha \sin \beta \cos c$$

Bibliography

ACIC (1959, December). Geodetic distance and azimuth computations for lines over 500 miles. Technical report 80, United States Air Force Aeronautical Chart and Information Center.

Adam, J., W. Augath, C. Boucher, C. Bruyninx, A. Caporali, E. Gubler, W. Gurtner, H. Habrich, B. Harsson, H. Hornik, J. Ihde, A. Kenyeres, H. van der Marel, H. Seeger, J. Simek, G. Stangl, J. Torres, and G. Weber (2002). Status of the European Reference Frame – EUREF. In J. Adam and K.-P. Schwarz (Eds.), *International Association of Geodesy Symposia*, Volume 125, pp. 42–46. IAG Scientific Assembly: Springer.

Alder, K. (2002). *The measure of all things*. New York: The Free Press.

Allister, N. and W. Featherstone (2001). Estimation of Helmert orthometric heights using digital barcode leveling, observed gravity and topographic mass-density data over part of Darling Scarp, Western Australia. *Geomatics Research Australia 75*, 25–52.

Altamimi, Z., X. Collilieux, J. Legrand, B. Garayt, and C. Boucher (2007). ITRF2005: A new release of the International Terrestrial Reference Frame based on time series of station positions and Earth Orientation Parameters. *Journal of Geophysical Research 112*(B09401), 1–19. doi:10.1029/2007JB004949.

Altamimi, Z., P. Sillard, and C. Boucher (2002). ITRF2000: A new release of the International Terrestrial Reference Frame for earth science applications. *Journal of Geophysical Research 107*(B10,2214), 2–1 – 2–19. doi:10.1029/2001JB000561.

Argus, D. and R. Gordon (1991). No-net-rotation model of current plate velocities incorporating plate motion model Nuvel-1. *Geophysical Research Letters 18*, 2038–2042.

Bazlov, Y. A., V. F. Galazin, B. L. Kaplan, V. G. Maksimov, and V. P. Rogozin (1999, July). Propagating PZ 90 to WGS 84 transformation parameters. *GPS Solutions 3*(1), 13–16.

Bearman, G. (1999). *Waves, tides and shallow-water processes* (2 ed.). Boston, Massachusetts: Butterworth-Heinemann.

Bermejo-Solera, M. and J. Otero (2009). Simple and highly accurate formulas for the computation of Transverse Mercator coordinates from longitude and isometric latitude. *Journal of Geodesy 83*(1), 1–12.

Berry, R. (1976). History of geodetic leveling in the United States. *Surveying and Mapping 36*(2), 137–153.

Blais, J. A. R., M. A. Chapman, and W. K. Lam (1986). Optimal interval sampling in theory and practice. In *Proceedings of the 2nd International Symposium on Spatial Data Handling*, Columbus, Ohio, pp. 185–192. International Geographical Union.

Blakely, R. J. (1995). *Potential theory in gravity and magnetic applications*. Cambridge: Cambridge University Press.

Boehm, W. and H. Prautzsch (1994). *Geometric concepts for geometric design*. Wellesley, Massachusetts: A K Peters.

Bomford, G. (1962). *Geodesy* (2 ed.). Oxford: Clarendon Press.

Boon, J. (2004). *Secrets of the tide: Tide and tidal current analysis and applications, storm surges and sea level trends*. Chichester, United Kingdom: Horwood Publishing.

Borkowski, K. M. (1989). Accurate algorithms to transform geocentric to geodetic coordinates. *Bulletin Géodésique 63*(1), 50–56.

Boucher, C. and Z. Altamimi (1986). Status of the realization of the BIH Terrestrial System. In A. K. Babcock and G. A. Wilkins (Eds.), *The Earth's Rotation and Reference Frames for Geodesy and Geodynamics: Proceedings of the 128th Symposium of the International Astronomical Union*, Symposium no. 128, Coolfont, West Virginia, U.S.A., pp. 20–24. International Astronomical Union: Kluwer Academic Publishers, Dordrecht.

Bowring, B. R. (1969, January). The further extension of the Gauss inverse problem. *Survey Review XX*(151), 40–43.

Bowring, B. R. (1971, July). The normal section – forward and inverse formulae at any distance. *Survey Review XX*(161), 131–135.

Bowring, B. R. (1972, April). Distance and the spheroid (correspondence). *Survey Review XXI*(164), 281–284.

Bowring, B. R. (1976, July). Transformation from spatial to geographical coordinates. *Survey Review XXIII*(181), 323–327.

Bowring, B. R. (1985). The accuracy of geodetic latitude and height equations. *Survey Review 28*(218), 202–206.

Bowring, B. R. (1987). Notes on the curvature in the prime vertical section. *Survey Review 29*(226), 195–196.

Bowring, B. R. (1996, July). Total inverse solutions for the geodesic and great elliptic. *Survey Review 33*(261), 461–476.

Broecker, W. (1983). The ocean. *Scientific American 249*, 79–89.

Brunner, F. K. (2002). The role of local quasi-dynamic heights in engineering geodesy. In *INGEO2002, 2nd Conference of Engineering Surveying*, Bratislava, pp. 21–24.

Bugayevskiy, L. M. and J. P. Snyder (1995). *Map projections: A reference manual*. Philadelphia, Pennsylvania: Taylor & Francis.

Buis, A. (2005). *NASA details earthquake effects on the Earth*. Jet Propulsion Laboratory. http://www.jpl.nasa.gov/news/news.cfm?release=2005-009.

Burša, M. (1969). Potential of the geoidal surface, the scale factor for lengths and earth's figure parameters from satellite observations. *Studia Geophysica et Geodaetica 13*, 337–358.

Burša, M. (1994). Testing geopotential models. *Earth, Moon and Planets 64*(3), 293–299.

Burša, M. (1995). *Report of Special Commission SC3, Fundamental constants*. Boulder, Colorado: The 21st General Assembly of the International Association of Geodesy. 2-14 July.

Burša, M., J. Kouba, M. Kumar, A. Müller, K. Radej, S. True, V. Vatrt, and M. Vojtíšková (1999). Geoidal geopotential and world height system. *Studia Geophysica et Geodaetica 43*(4), 327–337.

Burša, M., K. Radej, Z. Sima, S. True, and V. Vatrt (1997). Determination of the geopotential scale factor from topex/poseidon satellite altimetry. *Studia Geophysica et Geodaetica 41*(3), 203–216.

Campbell, W. H. (1997). *Introduction to geomagnetic fields*. New York: Cambridge University Press.

Casey, J. (1996). *Exploring curvature*. Braunschweig/Wiesbaden, Germany: Vieweg & Sohn Verlagsgesellschaft mbH.

Cazenave, A. and R. Nerem (2002). Redistributing earth's mass. *Science 297*(5582), 783–784.

Chao, B. and R. S. Gross (1987). Changes in the earth's rotation and low degree gravitational field induced by earthquakes. *Geophysics Journal of the Royal Astronomical Society 91*, 569–596.

Chelton, D., M. Schlax, M. Freilich, and R. Milliff (2004). Satellite measurements reveal persistent small-scale features in ocean winds. *Science 303*(5660), 978–982.

Clark, D. (1963). *Plane and geodetic surveying for engineers* (5 ed.), Volume 2. London: Constable and Co. Ltd.

Clarke, A. (1880). *Geodesy*. Oxford: Clarendon Press.

Cooperrider, A. Y., R. J. Boyd, and H. R. Stuart (Eds.) (1986). *Inventory and monitoring of wildlife habitat*. Denver, Colorado: U.S. Bureau of Land Management.

Crandall, C. L. (1914). *Geodesy and least squares*. New York: John Wiley & Sons.

Crane, N. (2003). *Mercator*. London: Orion Publishing Group, Ltd.

Danielsen, J. (1994). The use of Bessel-spheres for solution of problems related to geodesics on the ellipsoid. *Survey Review 32*(253), 445–449.

Davis, H. F. and A. D. Snider (1979). *Introduction to vector analysis* (4 ed.). Boston: Allyn and Bacon, Inc.

Dawson, J. J. and J. Steed (2004). International Terrestrial Reference Frame (ITRF) to GDA94 coordinate transformations. www.ga.gov.au/image_cache/GA3795.pdf.

Day, J. (1987). A refined chord-arc method of calculating geodesics. *Survey Review 29*(226), 191–194.

Dennis, M. L. and W. Featherstone (2002). Evaluation of orthometric and related height systems using a simulated mountain gravity field. In *3rd Meeting of the International Gravity and Geoid Commission: Gravity and Geoid 2002 - GG2002*, Section VI, Thessaloniki, Greece.

Densmore, D. and T. L. Heath (2000). *Euclid's elements*. Santa Fe, New Mexico: Green Lion Press.

Dewhurst, W. T. (1990). NADCON: The application of minimum-curvature-derived surface in the transformation of positional data from the North American Datum of 1927 to the North American Datum of 1983. Technical Report NOS NGS-50, National Oceanic and Atmospheric Administration, Rockville, Maryland.

Doodson, A. (1922). The harmonic development of the tide-generating potential. In *Proceedings of the Royal Society of London*, Series A 100, pp. 305–329.

Dracup, J. F. (1995). Geodetic surveys in the United States: The beginning and the next one hundred years 1807 - 1940. In *ACSM/ASPRS Annual Convention & Exposition Technical Papers*, Bethesda, Maryland, pp. 1–24.

Dubrovinsky, L. and J. F. Lin (2009). Mineral physics quest to the Earth's core. *EOS 90*(3), 21–22.

Elithorp Jr., J. A. and D. D. Findorff (2003). *Geodesy for geomatics and GIS professionals*. Ann Arbor, Michigan: XanEdu OriginalWorks.

Ewing, C. E. and M. Mitchell (1970). *Introduction to geodesy*. New York: American Elsevier Publishing Company, Inc.

Faller, J. E. and A. L. Vitouchkine (2003, May). Prospects for a truly portable absolute gravimeter. *Journal of Geodynamics 35*(4-5), 567–572.

Farin, G. (1993). *Curves and surfaces for computer aided geometric design: A practical guide*, Volume 3. New York: Academic Press.

Featherstone, W. and S. Claessens (2008). Closed-form transformation between geodetic and ellipsoidal coordinates. *Studia Geophysica et Geodaetica 52*(1), 1–18.

Featherstone, W. E. (1996). An updated explanation of the Geocentric Datum of Australia (GDA) and its effects upon future mapping. *The Australian Surveyor 41*(2), 121–130.

Feltens, J. (2008). Vector methods to compute azimuth, elevation, ellipsoidal normal, and the Cartesian (X, Y, Z) to geodetic (ϕ, λ, h) transformation. *Journal of Geodesy 82*(8), 493–504.

Fischer, I. (2004). Geodesy? What's that?, ch. 2. *ACSM Bulletin 208*, 43–52.

Forsberg, R. (1984). A study of terrain reductions, density anomalies and geophysical inversion methods in gravity field modeling. Technical Report Rep. 355, Department of Geodetic Science and Surveying, The Ohio State University, Columbus, Ohio.

Fukushima, T. (1999). Fast transform from geocentric to geodetic coordinates. *Journal of Geodesy 73*, 603–610.

Fukushima, T. (2006). Transformation from Cartesian to geodetic coordinates accelerated by Halley's method. *Journal of Geodesy 79*(12), 689–693.

Geeslin, N. and S. A. Brown (1989, April). With a campus legend in peril, members of a fraternity vow to save the endangered M.I.T. Smoot. *People Weekly 31*(16), 91.

Gerdan, G. and R. Deakin (1999). Transforming Cartesian coordinates X,Y,Z to geographical coordinates ϕ, λ, h. *Australian Surveying 44*(1), 55–63.

Gill, S. K. and J. R. Schultz (2001). Tidal datums and their applications. Technical Report Special Publication NOS CO-OPS 1, NOAA National Ocean Service, U.S. Coast and Geodetic Survey, Silver Spring, Maryland.

Glazmaier, G. A. and P. H. Roberts (1995, September). A three-dimensional self-consistent computer simulation of a geomagnetic field reversal. *Nature 377*, 203–209.

Glover, T. J. and R. A. Young (1996). *Measure for measure*. Littleton, Colorado: Sequoia Publishing, Inc.

Gore, J. H. (1889). *Elements of geodesy* (2 ed.). New York: John Wiley & Sons.

Grafarend, E. and A. Ardalan (1997). W0: An estimate in the Finnish Height Datum N60, epoch 1993.4, from twenty-five GPS points of the Baltic Sea Level Project. *Journal of Geodesy 71*(11), 673–679.

Groten, E. (2004). Fundamental parameters and current (2004) best estimates of the parameters of common relevance to astronomy, geodesy, and geodynamics. *Journal of Geodesy 77*(10-11), 724–731.

Hager, J. W., J. F. Behensky, and B. W. Drew (1989, September). The universal grids: Universal transverse mercator (utm) and universal polar stereographic (ups). Technical Report DMATM 8358.2, Defense Mapping Agency, Fairfax, Virginia.

Hall, S. (1992). *Mapping the next millennium*. New York: Random House.

Heikkinen, M. (1982). Geschlossene Formeln zur Berechnung räumlicher geodätischer Koordinaten aus rechtwinkligen Koordinaten. *Zeitschrift für Vermessungswesen 5*, 207–211.

Hein, G. (1985). Orthometric height determination using GPS observations and the integrated geodesy adjustment model. Technical Report 110 NGS 32, National Oceanic and Atmospheric Administration, Rockville, Maryland.

Heiskanen, W. A. and H. Moritz (1967). *Physical geodesy*. San Francisco: W. H. Freeman and Company.

Henning, W., E. Carlson, and D. Zilkoski (1998). Baltimore County, Maryland, NAVD 88 GPS-derived orthometric height project. *Surveying and Land Information Systems 58*(2), 97–113.

Hicks, S. (1985). Tidal datums and their uses: A summary. *Shore & Beach (Journal of the American Shore and Beach Preservation Association) 53*(1), 27–32.

Hicks, S. D. (2006). *Understanding tides*. Rockville, Maryland: U.S. Department of Commerce, NOAA.

Hipkin, R. (2002). Defining the geoid by $W = W_0 = U_0$: Theory and practice of a modern height system. In *Gravity and Geoid 2002, 3rd Meeting of the International Gravity and Geoid Commission*, Thessaloniki, Greece, pp. 367–377.

Hofmann-Wellenhof, B. H. and H. Moritz (2005). *Physical geodesy*. New York: SpringerWien-NewYork.

Hooijberg, M. (1997). *Practical Geodesy using computers*. Berlin, Germany: Springer.

Höpcke, W. (1966). On the curvature of electromagnetic waves and its effect on measurement of distance. *Survey Review 141*, 298–312.

Hothem, L. D., T. Vincenty, and R. E. Moose (1982). Relationship between Doppler and other advanced geodetic system measurements based on global data. In *Third International Geodetic Symposium on Satellite Doppler Positioning*, Las Cruces, New Mexico.

Hoyer, M., S. Arciniegas, K. Pereira, H. Fagard, R. Maturana, R. Torchetti, H. Drewes, M. Kumar, and G. Seeber (1998). The definition and realization of the reference system in the SIRGAS project. In J. Adam and K.-P. Schwarz (Eds.), *International Association of Geodesy Symposia*, Volume 118, pp. 167–173. IAG Scientific Assembly: Springer.

Hwang, C. (2002). Adjustment of relative gravity measurements using weighted and datum-free constraints. *Computers & Geosciences 28*(9), 1005–1015.

Hwang, C. and Y. Hsiao (2003). Orthometric corrections from leveling, gravity, density and elevation data: A case study in Taiwan. *Journal of Geodesy 77*(5-6), 279–291.

Ihde, J. and W. Augath (2000). The vertical reference system for Europe. In J. Torres and H. Hornik (Eds.), *Report on the Symposium of the IAG Subcommission for Europe (EUREF)*, pp. 99–101. June 22-24, Tromsö: Veröffentlichungen der Bayerischen Kommission für die Internationale Erdmessung, Astronomisch-Geodätische Arbeiten.

Iliffe, J. (2000). *Datums and map projections for remote sensing, GIS, and surveying.* New York: Whittles Publishing.

Ingle, J. C. J. (2000). Deep-sea and global ocean circulation. In W. G. Ernst (Ed.), *Earth systems: Processes and issues*, pp. 169–181. Cambridge, United Kingdom: Cambridge University Press.

Jank, W. and L. Kivioja (1980). Solution of the direct and inverse problems on reference ellipsoids by point-by-point integration using programmable pocket calculators. *Surveying and Mapping XL*, 325–337.

Jekeli, C. (2000). Heights, the geopotential, and vertical datums. Technical report 459, The Ohio State University.

Jekeli, C. (2005). *Geometric reference systems in geodesy.* Dept. of Geodetic Science and Surveying: The Ohio State University.

Jones, G. C. (2002). New solutions for the geodetic coordinate transformation. *Journal of Geodesy 76*(8), 437–446.

JPL (2004). Ocean surface topography from space. topex-www.jpl.nasa.gov/education/factor-height.html.

Kallay, M. (2007). Defining edges on a round earth. In *Proceedings of the 15th annual ACM international symposium on advances in geographic information systems*, Seattle, Washington.

Kao, S., R. Hsu, and F. Ning (2000). Results of field test for computing orthometric correction based on measured gravity. *Geomatics Research Australia 72*, 43–60.

Keay, J. (2000). *The great arc: The dramatic tale of how India was mapped and Everest was named.* New York: Harper Collins College Publishers.

Kellogg, O. D. (1953). *Foundations of potential theory.* New York: Dover Publications, Inc.

Kivioja, L. (1971, March). Computation of geodetic direct and indirect problems by computers accumulating increments from geodetic line elements. *Bulletin Géodésique 99*, 55–63.

Kovalevsky, J., I. I. Mueller, and B. Kolaczek (Eds.) (1989). *Reference frames in astronomy and geophysics.* Number 154 in Astrophysics and space science library. Boston: Kluwer Academic Publishers.

Kumar, M. (2005). When ellipsoidal heights will do the job, why look elsewhere? *Surveying and Land Information Science 65*(2), 91–94.

Laskowski, P. (1991). Is Newton's iteration faster than simple iteration for transformation between geocentric and geodetic coordinates? *Bulletin Géodésique 65*, 14–17.

Lee, L. P. (1944). The nomenclature and classification of map projections. *Empire Survey Review 7*, 190–200.

Lee, L. P. (1974, April). The computation of conformal projections. *Survey Review XXII*(172), 245–256.

Lee, L. P. (1983). Some conformal projections based on elliptic functions. *Geographical Review 55*(4), 563–580.

Lide, D. R. (Ed.) (2001). *CRC handbook of chemistry and physics* (81 ed.). Boca Raton, Florida: CRC Press, Inc.

Lin, K. and J. Wang (1995). Transformation from geocentric to geodetic coordinates using Newton's iteration. *Bulletin Géodésique 69*(4), 14–17.

Lindell, D. (2001). The problem corner. *Physical Science 21*(10), 34.

Linklater, A. (2002). *Measuring America*. New York: Walker Publishing Co.

Makarovic, B. (1973). Progressive sampling for digital terrain models. *ITC Journal 3*, 397–416.

Makarovic, B. (1977a). Composite sampling for digital terrain models. *ITC Journal 3*, 406–433.

Makarovic, B. (1977b). From progressive sampling to composite sampling for digital terrain models. *Geo–Processing 1*, 145–166.

Makarovic, B. (1984). Structures for geo–information and their application in selective sampling for digital terrain models. *ITC Journal 4*, 285–295.

Maling, D. (1973). *Coordinate systems and map projections*. London: George Philip & Son, Ltd.

Malys, S. and J. A. Slater (1994, November). Maintenance and enhancement of the World Geodetic System 1984. In *Proceedings of ION GPS-94*, Salt Lake City, Utah.

Maor, E. (1998). *Trigonometric delights*. Princeton, New Jersey: Princeton University Press.

Marmer, H. (1951). Tidal datum planes. Technical Report Special Publication No. 135, NOAA National Ocean Service, U.S. Coast and Geodetic Survey.

McCarthy, D. (1996, July). IERS conventions. Technical Note 21, IERS, Observatoire de Paris.

McCarthy, D. D. and G. Petit (Eds.) (2004). *IERS Conventions (2003)*, Volume IERS Technical Note 32. Frankfurt am Main: Verlag des Bundesamts für Kartographie und Geodäsie.

McCleary, J. (1994). *Geometry from a differentiable viewpoint*. Cambridge: Cambridge University Press.

McCullough, D. (1978). *Path between the seas: The creation of the Panama Canal, 1870-1914*. New York: Simon & Schuster.

McHenry, T. W. (2008). Fractional sections and the relationship of chains to acres. *The American Surveyor 5*(1), 60.

McLean, S., S. Macmillan, S. Maus, V. Lesur, A. Thomson, and D. Dater (2004, December). The US/UK World Magnetic Model for 2005-2010. NOAA Technical Report NESDIS/NGDC-1, National Oceanic and Atmospheric Administration.

Meeus, J. (1998). *Astronomical algorithms* (2 ed.). Richmond, Virginia: Willmann-Bell, Inc.

Melchior, P. (1974). Earth tides. *Geophysical Surveys 13*, 275–303.

Meyer, T. H. (2002, September). Grid, ground, and globe: Distances in the GPS era. *Surveying and Land Information Science 62*(3), 179–202.

Meyer, T. H., D. Roman, and D. B. Zilkoski (2005a). What does height really mean? Part I: Introduction. *Surveying and Land Information Science 64*(4), 223–234.

Meyer, T. H., D. Roman, and D. B. Zilkoski (2005b). What does height really mean? Part II: Physics and gravity. *Surveying and Land Information Science 65*(1), 5–15.

Meyer, T. H., D. Roman, and D. B. Zilkoski (2006a). What does height really mean? Part III: Height systems. *Surveying and Land Information Science 66*(2), 149–160.

Meyer, T. H., D. Roman, and D. B. Zilkoski (2006b). What does height really mean? Part IV: GPS orthometric heighting. *Surveying and Land Information Science 66*(3), 165–183.

Moffitt, F. H. and J. D. Bossler (1998). *Surveying* (10th ed.). Menlo Park, California: Addison-Wesley Publishing Company.

Molodensky, M. (1945). Fundamental problems of geodetic gravimetry (in Russian). Technical Report TRUDY Ts 42, Geodezizdat Novosibirskiy Institut Inzhenerov Geodezii, Aerofotos yemki i Kartografii (NIIGAiK), Moscow.

Moritz, H. (1980). *Advanced physical geodesy.* Tunbridge Wells, United Kingdom: Abacus Press.

Moritz, H. (2000). Geodetic reference system 1980. *Journal of Geodesy 74*(1), 128–162.

Mueller-Dombois, D. and H. Ellenberg (1974). *Aims and methods of vegetation ecology.* New York: John Wiley & Sons.

Murphy, D. W. (1981, January). Direct problem geodetic computation using a programmable pocket calculator. *Survey Review 26*(199), 11–16.

NADCON (1990). Notice to adopt standard method for mathematical horizontal datum transformation. *Federal Register 55*(155), 32681.

Nesvorny, D. and Z. Sima (1994). Refinement of the geopotential scale factor r0 on the satellite altimetry basis. *Earth, Moon and Planets 65*(1), 79–88.

NGA (2008). Daily 7-parameter Helmert transformation parameters between WGS 84(G1150) and ITRF 2000. earth-info.nga.mil/GandG/sathtml/index.html.

NGS (1981). *Geodetic leveling.* Rockville, Maryland: National Oceanic and Atmospheric Administration.

NGS (2001). DEFLEC99. www.ngs.noaa.gov/GEOID/DEFLEC99/.

NGS (2008). NAVD 88 gravity computations. www.ngs.noaa.gov/TOOLS/Navdgrav/navdgrav.shtml.

NGS (2009). Geodetic glossary. www.ngs.noaa.gov/CORS-Proxy/Glossary/xml/NGS_Glossary.xml.

NIMA (1997). Department of Defense World Geodetic System 1984: Its definition and relationships with local geodetic systems. Technical Report TR 8350.2, U.S. Defense Mapping Agency, Denver Federal Center: USGS Information Services.

NIST (2003). National Institute of Standards and Technology (NIST) reference on constants, units, and uncertainty. physics.nist.gov/cuu/Units/meter.html.

NOAA (2007). oceanservice.noaa.gov/education/kits/tides/media/supp_tide10b.html.

Pavlis, N., S. Holmes, S. Kenyon, and J. Factor (2008). An earth gravitational model to degree 2160: EGM2008. In *The 2008 General Assembly of the European Geosciences Union*, April 13-18, Vienna, Austria.

Pennycuick, C. J. (1988). *Conversion factors: SI units and many others*. Chicago: University of Chicago Press.

Peucker, T. K., R. J. Fowler, J. J. Little, and D. M. Mark (1978). The triangulated irregular network. In *Proceedings of the ASP Digital Terrain Models (DTM) Symposium*, Falls Church, Virginia, pp. 516–540. American Society of Photogrammetry.

Pollard, J. (2002). Iterative vector methods for computing geodetic latitude and height from rectangular coordinates. *Journal of Geodesy 76*(1), 36–40.

Pressley, A. (2007). *Elementary differential geometry*. London: Springer–Verlag.

Qihe, Y., J. P. Snyder, and W. R. Tobler (2000). *Map projection transformation principles and applications*. Philadelphia, Pennsylvania: Taylor & Francis.

Rainsford, H. F. (1949a). Long lines on the earth: Various formulæ. *Empire Survey Review 10*(71), 19–29.

Rainsford, H. F. (1949b). Long lines on the earth: Various formulæ. *Empire Survey Review 10*(72), 74–82.

Rainsford, H. F. (1955). Long geodesics on the ellipsoid. *Bulletin Géodésique 37*, 12–21.

Ramsey, A. (1981). *Newtonian attraction*. Cambridge, United Kingdom: Cambridge University Press.

Rapp, R. H. (1989a). *Geometric geodesy, part I*. Dept. of Geodetic Science and Surveying: The Ohio State University.

Rapp, R. H. (1989b). *Geometric geodesy, part II*. Dept. of Geodetic Science and Surveying: The Ohio State University.

Rebhan, H., M. Aguirre, and J. Johannessen (2000). The Gravity Field and Steady-State Ocean Circulation Explorer Mission – GOCE. *ESA Earth Observation Quarterly 66*, 6–11.

Robbins, A. R. (1962, July). Long lines on the spheroid. *Emperial Survey Review xvi*(125), 301–309.

Rollins, C. (2010, January). An integral for geodesic length after derivations by P. D. Thomas. *Survey Review 42*(315), 20–26.

Rudnicki, M. and T. H. Meyer (2007). Methods to convert local sampling coordinates into GIS/GPS-compatible coordinate systems. *Northern Journal of Applied Forestry 24*(3), 233–238.

Schey, H. M. (1992). *Div, grad, curl and all that: An informal text on vector calculus*. New York: W. W. Norton & Company.

Schwarz, C. R. (1989). North american datum of 1983. Professional Paper NOS 2, National Oceanic and Atmospheric Administration, Rockville, Maryland.

Shalowitz, A. (1938). The geographic datums of the coast and geodetic survey. *US Coast and Geodetic Survey Field Engineers Bulletin 12*, 10–31.

Sjöberg, Lars, E. (2007, January). Precise determination of the Clairaut constant in ellipsoidal geodesy. *Survey Review 39*(303), 81–86.

Sleep, N. H. and K. Fujita (1997). *Principles of geophysics*. Malden, Maryland: Blackwell Science.

Smith, D. A. (1998). There is no such thing as "The" EGM96 geoid: Subtle points on the use of a global geopotential model. *IGeS Bulletin 8*, 17–28.

Smith, D. A. and D. R. Roman (2001). GEOID99 and G99SSS:1-arc-minute geoid models for the United States. *Journal of Geodesy 75*(9), 469–490.

Smith, D. E. (1958). *History of mathematics: General survey of the history of elementary mathematics*, Volume 1. New York: Dover.

Smylie, D. and L. Manshina (1971). The elasticity theory of dislocation in real earth models and changes in the rotation of the earth. *Geophysics Journal of the Royal Astronomical Society 23*, 329–354.

Snay, R. and T. Soler (2000a, March). Reference systems: Part 3: WGS 84 and ITRS. *Professional Surveyor 20*(3), 1–3.

Snay, R. A. (1999). Using the HTDP software to transform spatial coordinates across time and between reference frames. *Surveying and Land Information Systems 59*(1), 15–25.

Snay, R. A. (2003). Introducing two spatial reference frames for regions of the Pacific Ocean. *Surveying and Land Information Science 63*(1), 5–12.

Snay, R. A. and T. Soler (2000b). Modern terrestrial reference systems part 2: The evolution of NAD 83. *Professional Surveyor 20*(2), 1–2.

Snyder, J. P. (1984). Minimum-error map projections bounded by polygons. *The Cartographic Journal 22*(22), 112–120.

Snyder, J. P. (1987). Map projections – a working manual. Professional Paper 1395, U.S. Geological Survey, Washington, D.C.

Sobel, D. and W. J. H. Andrewes (1995). *The illustrated longitude*. New York: Walker and Company.

Sodano, E. (1965). General non-iterative solution of the inverse and direct geodetic problems. *Bulletin Géodésique 75*, 69–89.

Soldati, G. and G. Spada (1999). Large earthquakes and earth rotation: The role of mantle relaxation. *Geophysical Research Letters 26*, 911–914.

Soler, T. (1998). A compendium of transformation formulas useful in GPS work. *Journal of Geodesy 72*(7-8), 482–490.

Soler, T. and J. Marshall (2002). Rigorous transformation of variance-covariance matrices of GPS-derived coordinates and velocities. *GPS Solutions 6*(1-2), 76–90.

Soler, T. and J. Marshall (2003). A note on frame transformations with applications to geodetic datums. *GPS Solutions 7*(1), 23–32.

Soler, T. and R. A. Snay (2004). Transforming positions and velocities between the International Terrestrial Reference Frame of 2000 and North American Datum of 1983. *Journal of Surveying Engineering 130*(2), 49–55.

Somiglinana, C. (1929). Teoria generale del campo gravitazionale dell'ellissoide di rotazione. *Memorie della Societa Astronomica Italiana IV*, 425.

Soycan, M. (2006). Determination of geoid heights by GPS and precise trigonometric levelling. *Survey Review 38*(299), 387–396.

Speed, Jr., M. F., H. J. Newton, and W. B. Smith (1996a, September). On the Utility of Mean Higher High Water and Mean Lower Low Water Datums for Texas inland coastal water. In *Oceans 96 MTS/IEEE*, Fort Lauderdale, Florida.

Speed, Jr., M. F., H. J. Newton, and W. B. Smith (1996b, June). Unfortunate law–unfortunate technology: The legal and statistical difficulties of determining coastal boundaries in areas with non-traditional tides. In *The Third International Conference on Forensic Statistics*, The University of Edinburgh, Scotland.

Stem, J. E. (1995, January). State plane coordinate system of 1983. NOAA Manual NOS NGS 5, National Oceanic and Atmospheric Administration, Rockville, Maryland.

Strang, G. and K. Borre (1997). *Linear algebra, geodesy, and GPS*. Wellesley, Massachusetts: Wellesley-Cambridge Press.

Strang van Hees, G. L. (1992). Practical formulas for the computation of the orthometric and dynamic correction. *Zeitschrift fur Vermessungswesen 117*, 727–734.

Taff, L. (1985). *Celestial mechanics. A computational guide for the practitioner*. New York: North–Holland.

Tapley, B., S. Bettadpur, M. Watkins, and C. H. Reigber (2004). The gravity recovery and climate experiment: Mission overview and early results. *Geophysical Research Letters 31*, L09607.

Taylor, A. (2004). *The world of Gerard Mercator: The mapmaker who revolutionized geography*. New York: Walker Publishing Company.

Tenzer, R., P. Vaníček, M. Santos, W. Featherstone, and M. Kuhn (2005). The rigorous determination of orthometric heights. *Journal of Geodesy 79*(1-3), 82–92.

Thien, G. (1967). *A solution to the inverse problem for nearly-antipodal points on the equator of the ellipsoid of revolution*. Master's thesis, The Ohio State University, Columbus, Ohio.

Thomas, P. (1970). Spheroidal geodesics, reference systems, & local geometry. Special publication 138, Naval Oceanographic Office, Washington, D.C.

Thomas, P. (1979). Conformal projections in geodesy and cartography. Special publication 251, Naval Oceanographic Office, Washington, D.C.

Torge, W. (1997). *Geodesy* (2 ed.). New York: Walter de Gruyter.

Vaníček, P. (1980). Tidal corrections to geodetic quantities. Technical Report NOS 83 NGS 14, National Oceanic and Atmospheric Administration, Rockville, Maryland.

Vaníček, P., J. Huang, P. Novak, S. Pagiatakis, M. Veronneau, Z. Martinec, and W. Featherstone (1999). Determination of the boundary values for the Stokes-Helmert problem. *Journal of Geodesy 73*(4), 180–192.

Vaníček, P. and E. Krakiwsky (1996). *Geodesy: The concepts* (3 ed.). Amsterdam: Elsevier Scientific Publishing Company.

Vassallo, A. and A. Secci (1995). Unique algorithm for the calculation of geodetic distances with the solution of the first fundamental geodetic problem. *Survey Review 33*(255), 50–58.

Vermeille, H. (2002). Direct transformation from geocentric coordinates to geodetic coordinates. *Journal of Geodesy 76*(8), 451–454.

Vincenty, T. (1971, July). The meridional distance problem for desk computers. *Survey Review XX*(161), 136–140.

Vincenty, T. (1975). Direct and inverse solutions of geodesics on the ellipsoid with application of nested equations. *Survey Review 33*(176), 88–93.

Volgyesi, L. (2006). Physical backgrounds of earth's rotation, revision of the terminology. *Acta Geodaetica et Geophysica Hungarica 41*(1), 31–44.

Weisstein, E. W. (2008). mathworld.wolfram.com/SphericalTriangle.html.

Weisstein, E. W. (2009). Coordinate system. mathworld.wolfram.com/CoordinateSystem.html. From MathWorld–A Wolfram Web Resource.

Welch, R. and A. Homsey (1997, April). Datum shifts for UTM coordinates. *Photogrammetric Engineering & Remote Sensing 63*(4), 371–375.

Whitehead, J. (1989). Giant ocean cataracts. *Scientific American 260*, 50–57.

Whitelaw, I. (2007). *A measure of all things*. New York: St. Martin's Press.

Wilhelm, H. and H.-G. Wenzel (1997). *Tidal phenomena*. New York: Springer.

Wolf, P. R. and B. A. Dewitt (2000). *Elements of photogrammetry with applications in GIS* (3 ed.). Boston: McGraw Hill.

Wolf, P. R. and C. D. Ghilani (1997). *Adjustment computations: Statistics and least squares in surveying and GIS*. New York: John Wiley & Sons.

Wolfram, S. (1999). *The Mathematica book* (4 ed.). Cambridge, United Kingdom: Cambridge University Press.

You, R.-J. (2000). Transformation of Cartesian to geodetic coordinates without iterations. *Journal of Surveying Engineering 126*(1), 1–7.

Zilkoski, D. (1990). Establishing vertical control using GPS satellite surveys. In *Proceedings of the 19th International Federation of Surveying Congress (FIG), Commission 5*, Helsinki, Finland, pp. 282–294.

Zilkoski, D. (2001). Vertical datums. In D. Maune (Ed.), *Digital elevation model technologies and applications: The DEM users manual*, pp. 35–60. Bethesda, Maryland: American Society for Photogrammetry and Remote Sensing.

Zilkoski, D. and L. Hothem (1989). GPS satellite surveys and vertical control. *Journal of Surveying Engineering 115*(2), 262–281.

Zilkoski, D., J. Richards, and G. Young (1992). Results of the general adjustment of the North American Vertical Datum of 1988. *Surveying and Land Information Systems 52*(3), 133–149.

Index

ϵ (first eccentricity), 65
ϵ' (second eccentricity), 65

algorithm, 11
angle, 22
    azimuth, 23
    bearing, 23
    dihedral, 137
    elevation, 22
    horizontal, 23
    spatial, 23
    spherical, 137
    vertical, 22
    zenith, 22
arc length, 103
arc-to-chord correction, 142
area, 26
    planimetric, 26
    slope, 26
axis
    semimajor, 57
    semiminor, 57
azimuth, 23
    grid, 32
    Laplace, 85
azimuth mark, 40

backsight, 34
bearing, 23
bench mark, 19
body tides, 218

central meridian, 145
circle, 13
    osculating, 66
circle (unit), 13
Clairaut's constant, 115
closure precision, 44
colatitude, 139

conformality, 153
contour line, 192
conventional reference pole, 137
convergence, 32, 142
coordinate
    scaling from a map, 29
coordinate system, 4, 73
    local, 5
coordinates, 4
    astronomic, 58, 132
    geographic, 58
    geometric, 58
    grid, 42, 160
    natural, 58
    survey control, 95
correction
    dynamic, 198
    free-air, 195
    normal, 200
    orthometric, 194
cosines, law of
    planar, 25
    spherical, 138
curve of alignment, 112

data sheets, 95
datum
    geodetic, 85
    NAVD 88, 204
    NGVD 29, 203
    U.S. National Vertical, 203
datums
    IGLD 85, 205
    tidal, 206
declination
    grid, 32, 142
    magnetic, 132
DEFLEC99, 134

deflection of the vertical, 133, 134, 181, 183
degrees
    decimal, 11
degrees of arc, 10
direct problem, 40
    spherical, 140
direction, 22
distance, 21
    definition, 104
    geodesic, 116, 117
    great circle, 108
        haversine, 109
    grid, 124
    ground, 105
    horizontal, 21, 105
    linear, 21
    slant, 21, 107
    slope, 21, 107
    spatial, 21, 107
    vertical, 21, 105
DMS (unit), 10
Doodson's constant, 216
dynamic correction, 198
dynamic height, 198
dynamic topography, 173

Earth mass, 174
Earth rotation parameters, 136
easting, 30, 31, 40, 42
eccentricity, 65
    first, 65
    second, 65
ecliptic, plane of, 136
elevation, 171
elevation factor, 124
ellipse
    oblate, 57
    prolate, 57
ellipsoid
    local, 185
    reference, 57
ephemeris, 129
epoch
    reference, 96
Equidistant cylindrical projection, 146

false easting, 162
false northing, 162
figure of the Earth, 181
flattening, 65
foot, 15
    international, 15
    survey, 15
foresight, 34
free-air correction, 195

geodesic
    ellipsoid, 116
geodesy, 3
    geometrical, 3
    physical, 3
geoid, 181
geoid height, 186
geoid undulation, 197
geopotential numbers, 188, 198
gon, 13
grad, 13
grade, 13
graticule, 145
gravitation, 174
gravity, 174
    normal, 175
great circle, 104, 108
great ellipse, 109
grid, 42
    map projection, 145

haversine formula, 109
height
    dynamic, 198
    ellipsoid, 76
    geoid, 123, 186, 197
    normal, 199
    orthometric, 191
        Helmert, 195
horizontal, 21

inversing, 28

Lambert Conformal Conic, 163
Laplace condition, 133
    extended, 133
Laplace correction, 133

INDEX

Laplace's equation, 133
latitude, 58
    astronomic, 132
    geocentric, 60
    geodetic, 67, 71
    reduced, 71
leveling, 38
    trigonometric, 38, 123
liftoff point, 116
linear closure error, 44
localization, 88
location, 4
longitude, 58
    astronomic, 132
    geocentric, 60
    geodetic, 63
loxodrome, 115, 152

map, 3
map projection systems, 164
map scale ratio, 28
marker
    survey, 19
mas (milli-arc second), 14
meridian, 60
    central, 145, 156
meter, 15
metes and bounds, 40
Metonic cycle, 136
mil, 13
    NATO, 13
    Soviet, 13
    US WWII, 13
milli-arc second, 14
minutes of arc, 10
monument, 19

NAD 27, 88, 166
NAD 83, 166
NADCON, 88
neat lines, 31
Newton's law of gravitation, 174
normal correction, 200
normal section, 69, 110
north, 129
    astronomic, 129, 132
    geodetic, 32, 129, 136
    grid, 32, 42, 129
    magnetic, 129
northing, 30, 31, 40, 42
nutation, 136, 211

ocean tides, 218
orientation, 40
orthometric correction, 194

parallel, 61
    central, 156
photogrammetry, 30, 81
planimetric, 26
plat, 5, 20, 106
Platonic year, 136
plumb line, 180, 191
point of beginning, 34, 40, 42
pointing, 22
position, 4, 42
    absolute, 19
    relative, 19
positioning, 42
    point, 129
potential energy, 180
precession, 136, 211
projection, 145
    aphylactic, 155
    aspect, 156
    azimuthal, 155
    conformal, 153, 155
    conic, 152
    cylindrical, 145
    direct aspect, 156
    equal-area, 155
    equidistant, 155
    Equidistant cylindrical, 146
    Lambert Conformal Conic, 163
    map, 145
    Mercator, 152, 158
    oblique aspect, 156
    planar, 149
    Plate Carrée, 146
    projection scale factor, 124
    standards, 126
    transverse aspect, 156

Transverse Mercator, 162

radian, 11
radius
 reference ellipsoid, average, 108
radius of curvature
 meridional, 67
 prime vertical, 68
reduction, 19
 horizontal to normal section
  rigorous, 123
  simplified, 124
reference frame
 alignment, 91
 Helmert transformation, 91
 scale, 90
 translation, 89
reference frames, 74
 conventional terrestrial, 74
 International Terrestrial, 74
reference system, 73
 conventional, 73
refraction, 107
representative fraction, 28
rhumb line, 115, 152
rod, 16
rubber sheeting, 99

scale, 28
 cartographic, 19
 determination, 30
second (time), 14
seconds of arc, 10
section, 69
 normal, 69, 110
sectorial constituent, 216
semicircle, 13
semicircle (unit), 13
sines, law of
 planar, 24
 spherical, 138
skew of the normals, 110
 formula, 110
south, 129
sphere
 authalic, 108

spherical excess, 162
surface
 developable, 145

tesseral constituent, 216
theodolite, 6
tide, 212
 body, 212
 earth, 212
total station, 20
transformation
 Helmert, 89
traverse, 20
triangle
 spherical, 137
triangulated irregular network, 105

universal gravitational constant, 174
Universal Transverse Mercator, 165

vertical, 21
 deflection of, 133
 local, 21

zonal constituent, 216
zone, 164

About the author

Thomas H. Meyer is an associate professor in the Department of Natural Resources and the Environment at the University of Connecticut. He received his doctorate from Texas A&M University, where he was a research associate in the Mapping Science Laboratory. Meyer is a member of the American Society of Civil Engineers, American Association for Geodetic Surveying, Cartography and Geographic Information Society, American Society for Photogrammetry and Remote Sensing, and American Geophysical Union. He has authored numerous peer-reviewed articles and papers about surveying and mapping. Additionally, Meyer teaches professional education seminars for surveyors throughout the United States.

Related titles from ESRI Press

Cartographic Relief Presentation

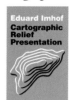

ISBN: 978-1-87910-026-1

Within the discipline of cartography, few works are considered classics in the sense of retaining their interest, relevance, and inspiration with the passage of time. One such work is Eduard Imhof's masterpiece, *Cartographic Relief Presentation*. Originally published in German in 1965, this book illustrates the need for cartography to combine intellect and graphics in solving map design problems. The range, detail, and scientific artistry of Imhof's solutions are presented in an instructional context that puts this work in a class by itself. ESRI Press has reissued Imhof's masterpiece as an affordable volume for mapping professionals, scholars, scientists, students, and anyone interested in cartography.

Lining Up Data: A Guide to Map Projections

ISBN: 978-1-58948-249-4

Data misalignment is one of the biggest issues faced by GIS users. As a member of the ESRI support services team, author Margaret Maher has handled more than 12,000 incidents relating to projections and data conversion. *Lining Up Data in ArcGIS* is a practical guide to solving these problems, based on Maher's decade of experience helping GIS users. This book presents techniques to identify data projections and create custom projections to align data. Formatted for practical use, each chapter can stand alone, addressing specific issues related to working with coordinate systems. *Lining Up Data in ArcGIS: A Guide to Map Projections* is a handbook that will benefit new and skilled GIS users alike.

Ocean Globe

ISBN: 978-1-58948-219-7

Ocean Globe focuses on bathymetry—the study of underwater depth of the third dimension—within the larger context of work being done by scientists and educators around the world. Each chapter represents a different facet of maritime research that relies on ocean floor mapping for its success. The topics covered address the diversity of the world's oceans and seas, placing emphasis on the need for better conservation. With a shared goal of joining disparate data collected over decades, the contributors of this volume turn to GIS as a tool for sharing information and advancing the science of bathymetry.

ESRI Press publishes books about the science, application, and technology of GIS. Ask for these titles at your local bookstore or order by calling 1-800-447-9778. You can also read book descriptions, read reviews, and shop online at www.esri.com/esripress. Outside the United States, contact your local ESRI distributor.